Quantitative Economics with R

Vikram Dayal

Quantitative Economics with R

A Data Science Approach

 Springer

Vikram Dayal
Indian Economic Service Section
Institute of Economic Growth
Delhi, India

ISBN 978-981-15-2037-2   ISBN 978-981-15-2035-8   (eBook)
https://doi.org/10.1007/978-981-15-2035-8

This Springer imprint is published by the registered company Springer Nature Singapore Pte Ltd.
The registered company address is: 152 Beach Road, #21-01/04 Gateway East, Singapore 189721, Singapore

For my old and wonderful friends
Ranu, Deepu and Chinna

Acknowledgements

The Institute of Economic Growth gave me a most conducive environment to work on the book. My colleagues have been supportive, two especially so: Prof. Bhavani and Purnamita. Suresh introduced me to R, and over the years shared my enthusiasm for R talk. Rahul provided inspiration with his painting about a sense of wonder, and specific advice on the text. Ranu edited most chapters and followed up on progress.

This book presents the work of a global community of scholars, coders and data scientists; not just economists like Acemoglu and Duflo, but statisticians like Paul Rosenbaum, the master of the tidyverse, Hadley Wickham, the deep philosopher of causal inference, Judea Pearl, the guru of networks, Barabasi, the experts in statistical learning, Hastie and Tibshirani. Writing this book led to a greater appreciation of their work and ideas.

Nupoor Singh at Springer and anonymous reviewers helped me with initial ideas for the book, and others at Springer facilitated the book very professionally.

My parents were encouraging. Varsha was unstinting in her support.

Vikram Dayal

Contents

About the Author

Vikram Dayal is a Professor at the Institute of Economic Growth, Delhi. He has been using the R software in teaching quantitative economics to diverse audiences, and is the author of the hugely popular SpringerBrief titled *An Introduction to R for Quantitative Economics: Graphing, Simulating and Computing.* Since its publication in 2015, each of its fourteen chapters has been downloaded at least 4000 times (and still counting) from SpringerLink. One can read the book at https://link.springer.com/book/10.1007%2F978-81-322-2340-5. He has published research on a range of environmental and developmental issues, from outdoor and indoor air pollution in Goa, India, to tigers and *Prosopis juliflora* in Ranthambore National Park. He studied economics in India and the USA, and received his doctoral degree from the Delhi School of Economics, University of Delhi.

Part I

Introduction to the Book and the Data Software

Introduction

1

1.1 A Data Science Approach

Why a data science approach to economics? Data science involves the intersection of computer science, statistics, which is a discipline that is about learning from data, and a knowledge domain—here, economics. In 2009, Hal Varian, then Google's chief economist, had provided the reasons for such an approach (McKinsey Quarterly 2009):

> The ability to take data—to be able to understand it, to process it, to extract value from it, to visualize it, to communicate it—that's going to be a hugely important skill in the next decades, not only at the professional level but even at the educational level for elementary school kids, for high school kids, for college kids. Because now we really do have essentially free and ubiquitous data. So the complementary scarce factor is the ability to understand that data and extract value from it.

> I think statisticians are part of it, but it's just a part. You also want to be able to visualize the data, communicate the data, and utilize it effectively. But I do think those skills—of being able to access, understand, and communicate the insights you get from data analysis—are going to be extremely important.

Note that Varian emphasizes that we need more than statistics or econometrics; we need to visualize and communicate.

In this book, we use the free software R (R Core Team 2019) to take a data science approach to quantitative economics. R was always amazing, Hadley Wickham has taken R to new levels by creating a set of packages, the `tidyverse`. Wickham (Grolemund and Wickham 2017) has thought carefully about the data analysis workflow, and he starts with the need to get the data (as does Varian). Then we spend a great deal of time 'wrangling' the data. We then graph the data to understand it, and fit models. We may also use graphs to communicate our analysis.

© Springer Nature Singapore Pte Ltd. 2020
V. Dayal, *Quantitative Economics with R*,
https://doi.org/10.1007/978-981-15-2035-8_1

1.2   Quick Tour of the Book

We take a quick tour of the book.

1.2.1   Part 1: Introduction to the Book and the Data Software

Part 1 of this book aims to provide just enough of an introduction to R. Though Wickham's work is well recognized, I think his contribution is crucial in helping the person who is new to R and has not written code. The use of what Wickham calls data verbs makes necessary wrangling of the data less of a chore; with time, it can be enjoyable. This part should convey some of the essential skills for getting the data in, working with the data and then making graphs with the data. There are 'Your turn' activities through the book that will build up skills.

1.2.2   Part 2: Managing and Graphing Data

One feature of data science that Wickham has greatly facilitated is wrangling and graphing. What Wickham grasped is that good visualization more often than not requires having suitably wrangled the data. When we graph data, we learn from data. Deaton (1997, pp. 3–4) explains his approach to analysing data:

> Rather than starting with the theory, I more often begin with the data and then try to find elementary procedures for describing them in a way that illuminates some aspect of theory or policy. Rather than use the theory to summarize the data through a set of structural parameters, it is sometimes more useful to present features of the data, often through simple descriptive statistics, or through graphical presentations of densities or regression functions, and then to think about whether these features tell us anything useful about the process whereby they were generated.

A few years back David Robinson suggested that we could teach R by taking newcomers through a direct tour of the tidyverse. We start with what has to be done: getting our data into R, before we enjoy the sheer pleasure of the bestselling ggplot2 package, Wickham's crowning achievement.

Modern data science deals with a rich variety of data, and an exciting type of data relates to networks. We start from a small toy network and contemplate the breathtaking complexity of global auto component trade networks.

1.2.3   Part 3: Mathematical Preliminaries for Data Analysis

We can use R for numerical mathematics, and we present some simple mathematics with R in Part 3. We can especially do a lot in R with difference equations.

1.2.4   Part 4: Inference from Data

R has for long been the *lingua franca* of statistics. Statistical and econometric teachers such as Kennedy (2003) have advocated the use of simulation, and it plays a central role in part 4 of the book, and in the first chapter of that part we use simulation to illuminate the central limit theorem. Two simulation-based inferential methods are presented: the bootstrap and randomization inference.

Simulation is also used to illuminate causal inference. We begin with a short look at causal graphs and potential outcomes, two frameworks that have greatly clarified issues in causal inference. We aim to understand and see examples of what Angrist and Pischke (2015) call the Furious Five—experiments, regression adjustment, regression discontinuity, difference-in-difference and instrumental variables. In addition, some applications of matching and a brief view of sensitivity analysis and Manski bounds are provided. As a result, the causal inference chapter is the longest in the book.

1.2.5   Part 5: Accessing, Analysing and Interpreting Growth Data

The next Part is devoted to economic growth. Data is visualized to examine the stylized facts of growth, and simple growth theory talks to the data. The second chapter in this Part looks at a key, famous paper by Acemoglu et al. (2001), their bold attempt at quantifying the effect of institutions on growth, and the mini-literature that developed around this paper.

1.2.6   Part 6: Basic Time Series Data

Part 6 deals with time series data. Time series graphs figure prominently in economics in the news. The first chapter deals with graphing time series.

The next chapter is a view of basic time series analysis. A section of this chapter uses simulation to illustrate key building blocks of time series models.

As in the rest of the book, in this part we consider it important to distinguish between description, prediction and causal inference.

1.2.7   Part 7: Introduction to Statistical/Machine Learning from Data

Varian (2014) wrote an engaging essay titled, *Big Data: New Tricks for Econometrics*. By now, the paper is old, and so are the tricks. We consider two key tricks of statistical learning: GAMS and random forests. Learning and using these and similar tricks is a big motivation to learn R, which is the software often used to implement such methods.

1.3   How to Use the Book

We can learn R in the same way we would learn a language. We should follow the book with RStudio open, typing in the R code and running it. We should experiment with the code, and see what happens. It is a good idea to use Google when we have doubts. Several **Your turn** activities are available through the book. Answers to some **Your turn** activities are at the end of the chapters. In this book, we usually start a topic with a small, easy example, so we can see what is going on. Selected resources for learning are mentioned at the end of chapters; these include online courses and videos.

If we are new to R or the tidyverse, we need to spend more time on the initial chapters.

1.4   Help

We can get help on a function in R by typing help followed by the function enclosed in parentheses; for example,

```
help(mean)
```

opens a help page on that function in RStudio.

Typing help.start() and running the command will open a page with hyperlinked manuals and package references in RStudio.

1.5   R Code and Output

In this book, the R code is in typewriter font. The resulting output is also indicated (with double hash) in typewriter font.

1.6   An Overview of Typical R Code

We can get lost in R code because there are so many commands and options; so we take a brief tour to get an overview. Typically, R code takes the form:

$$\boxed{\text{new object}} \leftarrow \boxed{\text{function}} (\boxed{\text{object or formula}}, \boxed{\text{object information}}, \boxed{\text{options}})$$

Not all the above elements come into a given line of code; what we have above is a generalization.

An example helps illustrate this more specifically:

```
• Price <- c(21, 31, 34)
```

This makes a vector called price; the c function is concatenate.

Installing and Loading Packages

Packages extend R's capabilities. We need to install a package once, before we can load it. We use the following code to install the tidyverse package:

```
install.packages("tidyverse")
```

We load the tidyverse package when we need to use it:

```
library(tidyverse)
```

Vectors

We can create a vector

```
Price <- c(2,3,4)
```

and get its third element with

```
Price[3]
```

Data

We can get a data file called myfile into R, and name it myfile:

```
myfile <- read.csv("myfile.csv")
```

we can access the second column with

```
second.column <- myfile[,2]
```

Graphs

We can draw a histogram of variable x with

```
library(tidyverse) # to load the tidyverse package
ggplot(mydata, aes(x = x)) +
  geom_histogram()
```

We can draw a scatter plot of y against x with

```
ggplot(mydata, aes(x = x, y = y)) +
  geom_point()
```

Regression

We can run a linear regression of y on x and z with

```
reg.mod <- lm(y ~ x + z, data = mydata)
```

To get regression output we can use:

```
summary(reg.mod)
```

1.7   Resources

Grolemund and Wickham (2017) is a beautiful book, to be kept on one's desk.

References

Acemoglu, D., S. Johnson, and J.A. Robinson. 2001. The colonial origins of comparative development: An empirical investigation. *American Economic Review* 91 (5): 1369–1401.
Angrist, J.D., and J. Pischke. 2015. *Mastering 'metrics*. Princeton: Princeton University Press.
Core Team, R. 2019. *R: A language and environment for statistical computing*. Vienna, Austria: R Foundation for Statistical Computing. https://www.R-project.org/.
Deaton, A. 1997. *The analysis of household surveys*. London: The Johns Hopkins University Press.
Grolemund, G., and H. Wickham. 2017. *R for data science*. Boston: O'Reilly.
Kennedy, P. 2003. *A guide to econometrics*, 5th ed. Cambridge: MIT Press.
Varian, H. 2009. How the web challenges managersTranscript of conversation with Manyika. McKinsey Quarterly. https://www.mckinsey.com/industries/technology-media-and-telecommunications/our-insights/hal-varian-on-how-the-web-challenges-managers [accessed 19 October 2019].
Varian, H. 2014. Big Data: New tricks for econometrics. *Journal of Economic Perspectives* 28 (2): 3–28.

RStudio and R

2

2.1 Introduction

R (R Core Team 2019) can be frustrating for learners. However, it can be enjoyable once you climb the learning curve. This chapter gets you started with some core features of R, and much of the book builds on these core features. So we see the same features being repeated, although with some elaboration.

2.2 R and RStudio

R is a highly flexible software. It is free. We can download it from:
http://www.r-project.org/.

In this book we work with R via RStudio, which makes our work easier. We can download RStudio (after installing R) from:
http://www.rstudio.com/.

If we experience any difficulty while downloading R or RStudio, we can simply use Google. For example, we could just search in Google for 'Installing R'. In general, using Google is a good idea when working with R.

Once we have installed R and RStudio, we only need to run RStudio.

Figure 2.1 shows a schematic of RStudio windows:

- Script or editor window. The top left window has an R script or alternative type of file, R Markdown, for example. We should always type our commands in an R script. By highlighting select code and clicking on run, we can run the selected lines of code.
- Console window. The bottom left window is the console window—this is where the output from R appears. There is a tab that says Console. We can type commands at the 'greater than' prompt, but it is better to use scripts.

© Springer Nature Singapore Pte Ltd. 2020
V. Dayal, *Quantitative Economics with R*,
https://doi.org/10.1007/978-981-15-2035-8_2

Fig. 2.1 Schematic of
RStudio windows

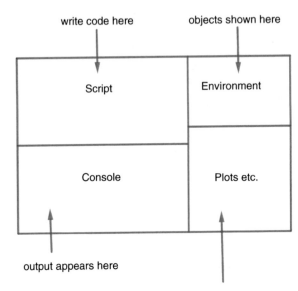

- The Environment or History window. The top right window has Environment and History tabs—different objects appear here as you create them. Under the Environment tab is 'Import Dataset', which we can use to import data into RStudio.
- Plots, etc. window. The bottom right window has the following tabs: Files, Plots, Packages, Help and Viewer. When graphs are made, they can be viewed here using the Plots tab. Packages can be installed with the Packages tab.

The four windows can be arranged depending on where we prefer to have them—top or bottom, right or left.

2.3   Use Projects

One of the most useful features of RStudio is the project facility. This helps a great deal with housekeeping; files and directories are arranged for us. We can create a new project by going to File, and then New Project. We can create a project and a new directory at the same time or we can create a new project in a directory. All output and files get saved in the same directory.

2.4   Use a Script

In this book, R code is in typewriter font. The resulting output is also indicated in typewriter font. The output line begins with a double hash. In R, we use functions that work with objects. In this book we will flag the use of a key function in the margins. In R we use packages that extend R; each package has a collection of functions. We will also flag packages.

We can start working with a script as follows. First, in RStudio we click on File, then New File, and then Script. We can save it as 'Script'. We can type in $2 + 3$, and click on Run; RStudio prints the result in the Console window. We can save the Script.

```
2 + 3
## [1] 5
```

There is an important subtlety. If the + sign is on the next line, R will think that the input finishes with 2; however, if the + is left after 2, R knows that some more input is on the next line:

```
2
## [1] 2
+ 3
## [1] 3
```

The code above is different from that below, because of where the + is placed.

```
# different placement of +
2 +
  3
## [1] 5
```

☞ **Your Turn**   Add 255 and 777 in R.

2.5   Different Objects in R

In R, we work with objects of different types. Let us use a simple example to briefly examine four important objects: vector, matrix, data frame and list.

2.5.1   Vectors

We set up a vector called Price, consisting of three prices. We need to type the following in the script window, and then click on Run, which runs that line. Then the line appears in the console window.

```
Price <- c(10,3,15)
```

The three prices are equal to 10, 3 and 15. We use the c() function which stands for concatenate, and parentheses enclose the values that are separated by commas.

When we run the command above, we don't see any output. R simply creates the object called Price, and you can see it in the Environment window. To print it, we need to type Price and run the line:

```
Price
## [1] 10  3 15
```

We notice that the output includes [1]; this only tells us that the first element is ten.

R will distinguish between Price and price; if we are not careful we get an error message.

```
price

## Error in eval(expr, envir, enclos): object 'price' not found
```

In R, a parenthesis ( ) is different from a bracket [ ]—each has to be used in the right way depending on the context.

```
Price <- c[10,3,15]

## Error in c[10, 3, 15]: object of type 'builtin' is not
subsettable
```

We can find out the length of our vector Price:

```
length(Price)
## [1] 3
```

We can extract the first element:

```
Price[1]
## [1] 10
```

and the second and third elements:

```
Price[2:3]
## [1]  3 15
```

We create a vector for corresponding quantities and print it:

```
Quantity <- c(25,3,20)
Quantity
## [1] 25  3 20
```

We can multiply the Price and Quantity vectors, which gives us Expenditure.

```
Expenditure <- Price * Quantity
Expenditure
## [1] 250    9 300
```

The sum of the elements of Expenditure gives us total Expenditure.

```
Total_expenditure <- sum(Expenditure)
Total_expenditure
## [1] 559
```

Getting comfortable with vectors is very important to learn R.

> ☞ **Your Turn** In an R Script, type in the following two vectors in R: $Price2 \leftarrow c(2, 12, 7, 33)$ and $Quantity2 \leftarrow c(23, 32, 12, 33)$. Then extract the fourth element of Price2 and Quantity2, calculate Expenditure2 and TotalExpenditure2, following what we did above with the vectors Price and Quantity. Save the script file, we will continue the exercise.

2.5.2 Matrices

The Price, Quantity and Expenditure vectors can be bound into the columns of a matrix using the `matrix` function:

```
Matrix_PQE <- matrix(data = cbind(Price, Quantity, Expenditure)
   , ncol=3)
Matrix_PQE
##       [,1] [,2] [,3]
## [1,]   10   25  250
## [2,]    3    3    9
## [3,]   15   20  300
```

We used the R function `matrix` above and also the function `cbind`, which binds the vectors into columns.

We print the first row of the matrix.

```
Matrix_PQE[1,]
## [1]  10  25 250
```

and then the second column.

```
Matrix_PQE[,2]
## [1] 25  3 20
```

First row, second column:

```
Matrix_PQE[1,2]
## [1] 25
```

The first number between the brackets indicates the row, the second the column. We discuss matrices in R in a later chapter.

2.5.3   Data Frames

We can create a data frame and print it:

```
Exp_data <- data.frame(Price, Quantity)
Exp_data
##    Price Quantity
## 1     10       25
## 2      3        3
## 3     15       20
```

We print the second column.

```
Exp_data[,2]
## [1] 25  3 20
```

We can also refer to the second column of the data frame by using a dollar sign and the name of the column:

```
Exp_data$Quantity
## [1] 25  3 20
```

We discuss getting data into R in the next chapter.

☞ **Your Turn**   Create a data frame containing the vectors Price2 and Quantity2. Try to print the first column of the data frame.

2.5.4   Lists

A list is a collection of heterogeneous objects. We create a list containing some of the expenditure objects we have created, with the list() function.

```
Expenditure_list <- list(Price, Quantity, Expenditure,
  Total_expenditure)
Expenditure_list
```

```
## [[1]]
## [1] 10   3 15
##
## [[2]]
## [1] 25   3 20
##
## [[3]]
## [1] 250    9 300
##
## [[4]]
## [1] 559
```

The index for a list uses a double bracket. We print the second element below.

```
Expenditure_list[[2]]
## [1] 25   3 20
```

2.6   Toy Example: Net Present Value

We calculate the present value of a sum of money (121) received two years from now, the discount rate is 10%. First, we tell R what the values are:

```
Amount <- 121
discount_rate <- 0.10
time <- 2
```

Then we tell R how to calculate the net present value.

```
Net_present_value <- Amount/(1+discount_rate)^time
Net_present_value
## [1] 100
```

Another example. We now calculate the net present value of several sums of money. A cost of 150 is incurred now, and benefits of 135 and 140 are received after one and two years. The discount rate continues to be 10%. We use the concatenate (i.e. c()) function.

```
Cost_benefit_profile <- c(-150,135,140)
time_profile <- c(0,1,2)
```

We give R the formula for the profile of discounted costs and benefits.

```
Cost_benefit_present_value_profile <-
  Cost_benefit_profile/(1+discount_rate)^time_profile
```

We sum the values and print; we can use the round function to round off.

```
Net_present_value   <- sum(Cost_benefit_present_value_profile)
Net_present_value
## [1] 88.42975
round(Net_present_value, digits = 0)
## [1] 88
```

We need to be careful while working with vectors, paying attention to their dimensions. Below, we add a vector *Three* with three elements to a vector *Two* with one element.

```
Three <- c(3,3,3)
Two <- 2
Five <- Three + Two

Five
## [1] 5 5 5
```

What if we add the vector *Three* with three elements to a vector *Mix* with two elements?

```
Mix <- c(2,9)
Mix
## [1] 2 9
ThreeandMix <- Three + Mix

## Warning in Three + Mix: longer object length is not
a multiple of shorter object length

ThreeandMix
## [1]  5 12  5
```

After issuing the warning, R 'recycles' *Mix*; since the third element is missing it goes back to the first.

2.7   The Tidyverse Approach

We now present Hadley Wickham's approach to the data analysis workflow. This approach is exposited in Grolemund and Wickham (2017). Wickham has designed several packages to work on different parts of a workflow.

2.7.1   Data Analysis Workflow

Typically, a data analyst follows a workflow consisting of these parts:

1. Get data into R,
2. tidy and transform,
3. visualize and model, and
4. communicate.

By working systematically, we can make steady progress and be more efficient. Working systematically involves using script files and the project feature in RStudio. Script files have a .R extension. When we use the project feature in RStudio, it is easier for different files to work together. In addition to documenting commands in script files, we will use files for data (often csv files) and produce figures or documents that can be pdf or word files, for instance.

Now we will use hypothetical data. This toy example helps us get into a data analysis workflow. Because it is a toy example with a small set of data, it is easy to see what is going on.

If data is tidy it makes analysis easier. So what is a tidy dataset? Briefly, all the variables are in columns and all the observations are in rows.

2.7.2   The Tidyverse Package

The 'tidyverse' (Wickham 2017) is a set of packages that have been written by Hadley Wickham that help us with the different aspects of data analysis workflow. We will now use Wickham's `tidyverse` package. This requires us to first install the package. A package has to be installed only once. It can be installed either with a command or via an icon in RStudio. Subsequently, we can load the package with the library function.

```
# is a comment
# install once
# install.packages("tidyverse")
# load each time

library(tidyverse)
```

2.7.3   Input and Wrangle Synthetic Data

We enter our data in R with each variable as a separate vector. This is hypothetical data of a survey of six persons, and the variables are: payment they received, hours worked, their gender and age.

```
surv_id <- c(1,2,3,4,5,6)
payment <- c(1000,700,600,1200,800,500)
hours <- c(7,5,3,6,7,4)
gender <- c("F","M","F","M","M","M")
age <- c(28,52,37,35,59,43)
```

In R, data is usually stored in a data frame. Wickham has designed a 'tibble' to improve on R's data frame. A tibble, which is similar to a data frame, stores our data. We call our tibble labour. We create a tibble with the `tibble()` function.

```
labour <- tibble(surv_id,
  payment, hours,
  gender,age)
```

We print labour. Note that R would otherwise only create the tibble, and would do so quietly.

```
labour
## # A tibble: 6 x 5
##    surv_id payment hours gender   age
##      <dbl>   <dbl> <dbl> <chr>  <dbl>
## 1        1    1000     7 F         28
## 2        2     700     5 M         52
## 3        3     600     3 F         37
## 4        4    1200     6 M         35
## 5        5     800     7 M         59
## 6        6     500     4 M         43
```

We get a glimpse of our data with the `glimpse` function.

```
glimpse(labour)
## Observations: 6
## Variables: 5
## $ surv_id <dbl> 1, 2, 3, 4, 5, 6
## $ payment <dbl> 1000, 700, 600, 1200, 800, 500
## $ hours   <dbl> 7, 5, 3, 6, 7, 4
## $ gender  <chr> "F", "M", "F", "M", "M", "M"
## $ age     <dbl> 28, 52, 37, 35, 59, 43
```

We can now write our data to a csv file, with the `write_csv` function, and then we read this file into RStudio with the `read_csv` function:

```
write_csv(labour, "labour.csv")
labour2 <- read_csv("~/Documents/R/ies2018/labour.csv")

## Parsed with column specification:
## cols(
##   surv_id = col_double(),
##   payment = col_double(),
##   hours = col_double(),
##   gender = col_character(),
##   age = col_double()
## )
```

We see that the tibble named labour2 is identical to the tibble named labour.

```
labour2
## # A tibble: 6 x 5
##    surv_id payment hours gender   age
##      <dbl>   <dbl> <dbl> <chr>  <dbl>
## 1        1    1000     7 F         28
## 2        2     700     5 M         52
## 3        3     600     3 F         37
## 4        4    1200     6 M         35
## 5        5     800     7 M         59
## 6        6     500     4 M         43
```

☞ **Your Turn** Using an R script, create tibble labour as above, then write it as a csv file, as above, and read it in as above. The tibble labour2 will be used ahead, so save the R script.

A tibble consists of several vectors. We extract the gender column from the labour dataset, and then the second and third elements of the gender column:

```
labour$gender
## [1] "F" "M" "F" "M" "M" "M"
labour$gender[2:3]
## [1] "M" "F"
```

We extract the first row and then the second column.

```
labour[1,]
## # A tibble: 1 x 5
##    surv_id payment hours gender   age
##      <dbl>   <dbl> <dbl> <chr>  <dbl>
## 1        1    1000     7 F         28
labour[,2]
## # A tibble: 6 x 1
##    payment
##      <dbl>
## 1     1000
## 2      700
## 3      600
## 4     1200
## 5      800
## 6      500
```

2.7.4   Five Data Verbs

Five data verbs help us do a lot with data. We use these with the help of the `dplyr` package written by Wickham, contained in the `tidyverse` package.

The five data verbs are:

filter	to pick certain rows
select	to pick columns
mutate	to generate new variables
summarize	to summarize
arrange	to sort in some order

All of these can be used with the `group_by()` function.

These five verbs, along with the pipe symbol, help us accomplish a lot when working with data. The pipe symbol

```
%>%
```

helps us string together commands. The following commands are equivalent.

```
f(x,y)
```

is equivalent to

```
x %>%
f(,y)
```

The pipe symbol pipes x into the function of x and y. So what is on the left of the pipe gets piped as the first argument of the function on the right. This might appear strange at first and takes some getting used to, but it greatly helps us carry out actions on data that build on each other and makes code easier to understand.

We select rows by using `filter`. We filter for rows where the gender is female. We create a new tibble, by piping labour into the filter function, asking it to give us the rows with female gender (F). Note the use of two equal to signs.

```
labour_filter <- labour %>%
  filter(gender == "F")

labour_filter
## # A tibble: 2 x 5
##   surv_id payment hours gender   age
##     <dbl>   <dbl> <dbl> <chr>  <dbl>
## 1       1    1000     7 F         28
## 2       3     600     3 F         37
```

We could have avoided using the pipe symbol as below, but there are advantages to using it.

```
labour_filter <- filter(labour, gender == "F")

labour_filter
## # A tibble: 2 x 5
##    surv_id payment hours gender   age
##      <dbl>   <dbl> <dbl> <chr>  <dbl>
## 1        1    1000     7 F         28
## 2        3     600     3 F         37
```

> ☞ **Your Turn**   Filter for rows where the gender is male, calling the new tibble labour_filter2. Check that you have done this correctly by printing labour_filter2.

We create new variables with mutate; we calculate the wage rates.

```
labour_mutate <- labour %>%
  mutate(wage = payment /
           hours)

labour_mutate
## # A tibble: 6 x 6
##    surv_id payment hours gender   age  wage
##      <dbl>   <dbl> <dbl> <chr>  <dbl> <dbl>
## 1        1    1000     7 F         28  143.
## 2        2     700     5 M         52  140
## 3        3     600     3 F         37  200
## 4        4    1200     6 M         35  200
## 5        5     800     7 M         59  114.
## 6        6     500     4 M         43  125
```

> ☞ **Your Turn**   Create a new variable called minutes with mutate: minutes = hours * 60.

We arrange the data by hours worked with arrange.

```
labour_arrange <- labour %>%
  arrange(hours)
```

```
labour_arrange
## # A tibble: 6 x 5
##    surv_id payment hours gender    age
##      <dbl>   <dbl> <dbl> <chr>   <dbl>
## 1        3     600     3 F          37
## 2        6     500     4 M          43
## 3        2     700     5 M          52
## 4        4    1200     6 M          35
## 5        1    1000     7 F          28
## 6        5     800     7 M          59
```

We select the columns hours worked and gender.

```
labour_select <- labour %>%
  select(hours, gender)

labour_select
## # A tibble: 6 x 2
##   hours gender
##   <dbl> <chr>
## 1     7 F
## 2     5 M
## 3     3 F
## 4     6 M
## 5     7 M
## 6     4 M
```

We now summarize the data; grouping by gender. The group by here groups by gender; we get the mean hours worked by females and males.

```
labour_summary <- labour %>%
  group_by(gender) %>%
  summarize(mean = mean(hours))

labour_summary
## # A tibble: 2 x 2
##   gender   mean
##   <chr>   <dbl>
## 1 F           5
## 2 M         5.5
```

☞ **Your Turn**   Find the median hours worked by gender.

Fig. 2.2 gg1 gives us axes only

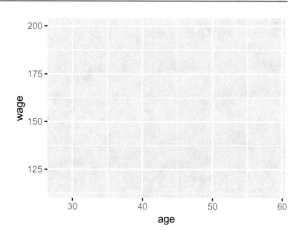

2.7.5   Graphs

The package ggplot2 in the tidyverse can do wonderful visualizations. In ggplot2, gg stands for the grammar of graphics. Here we ask for a plot; we mention the x and y variable as aesthetics that are mapped onto the axes in Fig. 2.2.

```
gg1 <- ggplot(data = labour_mutate,
         aes(x = age, y = wage))
gg1


# see Figure 2.2
```

We use geom_point() to tell R that we want points plotted. The different components are added with a plus at the end of the line of code, and this helps us build plots in layers (Fig. 2.3).

```
gg2 <- gg1 +
  geom_point()
gg2


# see Figure 2.3
```

The colour aesthetic is used to distinguish the gender of the cases (Fig. 2.4).

```
gg3 <- gg1 +
  geom_point(
    aes(colour = gender))
gg3


# see Figure 2.4
```

We get separate graphs with facet_wrap and are able to distinguish between males and females (Fig. 2.5).

Fig. 2.3 gg2 gives us a
scatter plot

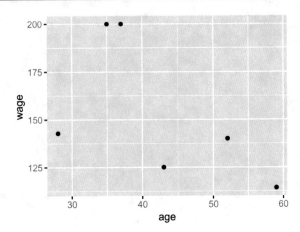

Fig. 2.4 gg3 gives us
markers with colour
denoting gender

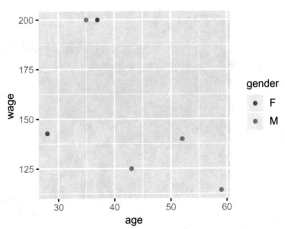

```
gg4 <- gg2 +
  facet_wrap(~ gender)
gg4
```

```
# see Figure 2.5
```

2.7.6   Linear Model

We can fit a linear model with the lm command. This creates an object which we
can print or extract information from.

```
age_wage_fit <- lm(wage ~ age, data = labour_mutate)

age_wage_fit
##
## Call:
```

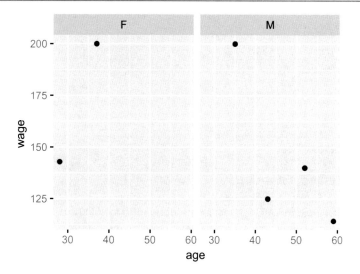

Fig. 2.5 gg4 gives us a scatter plot faceted by gender

```
## lm(formula = wage ~ age, data = labour_mutate)
##
## Coefficients:
## (Intercept)          age
##      233.28        -1.88
#can use summary(age_wage_fit)
```

Our fitted equation is $wage = 233 - 1.88\,age$.

We can plot the fitted line along with the scatter plot. We add geom_smooth(), and ask for it to be based on a linear model (Fig. 2.6).

Fig. 2.6 gg5 gives us a
scatter plot with a fitted line

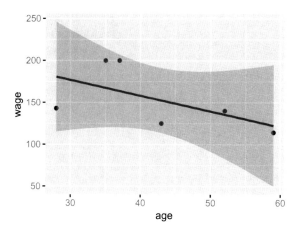

```
gg5 <- gg2 +
  geom_smooth(method = "lm")
gg5
```

```
# see Figure 2.6
```

2.8   Resources

For Better Understanding

The online Datacamp course (https://www.datacamp.com/courses/free-introduction-to-r), Introduction to R, by Jonathan Cornelissen, is very useful if you are new to R.

The online Datacamp course (https://www.datacamp.com/courses/introduction-to-the-tidyverse), Introduction to the Tidyverse, by David Robinson, is very useful if you are new to the tidyverse.

For Going Further

As already mentioned in an earlier chapter, R for Data Science by Grolemund and Wickham (2017) is the cutting edge book on R, useful to keep on one's desk.

Select Your Turn Answers

> ☞ **Your Turn (Sect. 2.5)**   In an R Script, type in the following two vectors in R: $Price2 \leftarrow c(2, 12, 7, 33)$ and $Quantity2 \leftarrow c(23, 32, 12, 33)$. Then extract the fourth element of Price2 and Quantity2, and calculate Expenditure2 and TotalExpenditure2, following what we did above with the vectors Price and Quantity. Save the script file, and we will continue the exercise.

```
Price2 <- c(2, 12, 7, 33)
Quantity2 <- c(23, 32, 12, 33)
Price2[4]
## [1] 33
Quantity2[4]
## [1] 33
```

```
Expenditure2 <- Price2 * Quantity2
Expenditure2
## [1]   46  384    84 1089
TotalExpenditure2 <- sum(Expenditure2)
TotalExpenditure2
## [1] 1603
```

☞ **Your Turn (Sect. 2.7)**   Filter for rows where the gender is male, calling the new tibble labour_filter2. Check that you have done this correctly by printing labour_filter2.

```
labour_filter2 <- filter(labour, gender == "M")
labour_filter2
## # A tibble: 4 x 5
##   surv_id payment hours gender   age
##     <dbl>   <dbl> <dbl> <chr>   <dbl>
## 1       2     700     5 M          52
## 2       4    1200     6 M          35
## 3       5     800     7 M          59
## 4       6     500     4 M          43
```

References

Core Team, R. 2019. *R: A language and environment for statistical computing*. Vienna, Austria: R Foundation for Statistical Computing. https://www.R-project.org/.

Grolemund, G., and H. Wickham. 2017. *R for data science*. Boston: O'Reilly.

Wickham, H. 2017. tidyverse: Easily Install and Load 'Tidyverse' Packages. *R Package Version* 1(1): 1. https://CRAN.R-project.org/package=tidyverse.

Part II
Managing and Graphing Data

Getting Data into R

3

3.1 Introduction

We will look at a few ways to get your data into R; while not exhaustive, they orient us to this important step in data analysis.

3.2 Data in R or a Package

Some data is available in R itself. One example is the `anscombe` data. We use the `data()` function.

```
data(anscombe)
```

We call the data `ans` to make typing easier.

```
ans <- anscombe
```

We can use the `str()` function to see the structure of the data.

```
str(ans)
## 'data.frame':   11 obs. of  8 variables:
##  $ x1: num  10 8 13 9 11 14 6 4 12 7 ...
##  $ x2: num  10 8 13 9 11 14 6 4 12 7 ...
##  $ x3: num  10 8 13 9 11 14 6 4 12 7 ...
##  $ x4: num  8 8 8 8 8 8 8 19 8 8 ...
##  $ y1: num  8.04 6.95 7.58 8.81 8.33 ...
##  $ y2: num  9.14 8.14 8.74 8.77 9.26 8.1 6.13 3.1 9.13 7.26 ...
##  $ y3: num  7.46 6.77 12.74 7.11 7.81 ...
##  $ y4: num  6.58 5.76 7.71 8.84 8.47 7.04 5.25 12.5 5.56 7.91 ...
```

© Springer Nature Singapore Pte Ltd. 2020
V. Dayal, *Quantitative Economics with R*,
https://doi.org/10.1007/978-981-15-2035-8_3

We can see the top rows of ans:

```
head(ans)
##    x1 x2 x3 x4   y1   y2    y3   y4
## 1 10 10 10   8 8.04 9.14  7.46 6.58
## 2  8  8  8   8 6.95 8.14  6.77 5.76
## 3 13 13 13   8 7.58 8.74 12.74 7.71
## 4  9  9  9   8 8.81 8.77  7.11 8.84
## 5 11 11 11   8 8.33 9.26  7.81 8.47
## 6 14 14 14   8 9.96 8.10  8.84 7.04
```

We now look at data available in a package. We need to install and load the
package first. Here we use the wooldridge package (Shea 2018).

```
library(wooldridge)
data(injury)
str(injury)
## 'data.frame':    7150 obs. of   30 variables:
##  $ durat   : num  1 1 84 4 1 1 7 2 175 60 ...
##  $ afchnge : int  1 1 1 1 1 1 1 1 1 1 ...
##  $ highearn: int  1 1 1 1 1 1 1 1 1 1 ...
##  $ male    : int  1 1 1 1 1 1 1 1 1 1 ...
##  $ married : int  0 1 1 1 1 1 1 1 1 1 ...
##  $ hosp    : int  1 0 1 1 0 0 0 1 1 1 ...
##  $ indust  : int  3 3 3 3 3 3 3 3 3 3 ...
##  $ injtype : int  1 1 1 1 1 1 1 1 1 1 ...
##  $ age     : int  26 31 37 31 23 34 35 45 41 33 ...
##  $ prewage : num  405 644 398 528 529 ...
##  $ totmed  : num  1188 361 8964 1100 373 ...
##  $ injdes  : int  1010 1404 1032 1940 1940 1425 1110 1207 1425
##                   1010 ...
##  $ benefit : num  247 247 247 247 212 ...
##  $ ky      : int  1 1 1 1 1 1 1 1 1 1 ...
##  $ mi      : int  0 0 0 0 0 0 0 0 0 0 ...
##  $ ldurat  : num  0 0 4.43 1.39 0 ...
##  $ afhigh  : int  1 1 1 1 1 1 1 1 1 1 ...
##  $ lprewage: num  6 6.47 5.99 6.27 6.27 ...
##  $ lage    : num  3.26 3.43 3.61 3.43 3.14 ...
##  $ ltotmed : num  7.08 5.89 9.1 7 5.92 ...
##  $ head    : int  1 1 1 1 1 1 1 1 1 1 ...
##  $ neck    : int  0 0 0 0 0 0 0 0 0 0 ...
##  $ upextr  : int  0 0 0 0 0 0 0 0 0 0 ...
##  $ trunk   : int  0 0 0 0 0 0 0 0 0 0 ...
##  $ lowback : int  0 0 0 0 0 0 0 0 0 0 ...
##  $ lowextr : int  0 0 0 0 0 0 0 0 0 0 ...
##  $ occdis  : int  0 0 0 0 0 0 0 0 0 0 ...
##  $ manuf   : int  0 0 0 0 0 0 0 0 0 0 ...
##  $ construc: int  0 0 0 0 0 0 0 0 0 0 ...
##  $ highlpre: num  6 6.47 5.99 6.27 6.27 ...
##  - attr(*, "time.stamp")= chr "25 Jun 2011 23:03"
```

If we want just the names of the variables, we could use:

```
names(injury)
##   [1] "durat"     "afchnge"   "highearn"  "male"
##   [5] "married"   "hosp"      "indust"    "injtype"
##   [9] "age"       "prewage"   "totmed"    "injdes"
##  [13] "benefit"   "ky"        "mi"        "ldurat"
##  [17] "afhigh"    "lprewage"  "lage"      "ltotmed"
##  [21] "head"      "neck"      "upextr"    "trunk"
##  [25] "lowback"   "lowextr"   "occdis"    "manuf"
##  [29] "construc"  "highlpre"
```

3.3   Data in a csv File

The Chhatre and Agrawal (2009) data is available at: http://ifri.forgov.org/resources/data/.

We go to the Referenced Datasets section of the website, and the data corresponding to Chhatre and Agrawal's (2009) paper, "Carbon storage and livelihoods". I had last accessed it on 4 October 2019. It is a zip file and contains a csv (comma-separated variable) file.

After downloading it to the same directory as our current project, we go to the environment window and click on Import Dataset; then from text (readr) (Wickham et al. 2018) we choose File "ifri_car_liv", and select header. Alternatively, we can type in the following command, changing the file path as required. We use underscores in our name for the dataset to avoid blank spaces.

```
library(readr)
ifri_car_liv <- read_csv("ifri_car_liv.csv")

## Parsed with column specification:
## cols(
##   forest_id = col_double(),
##   cid = col_character(),
##   zliv = col_double(),
##   zbio = col_double(),
##   livcarl = col_double(),
##   ownstate = col_double(),
##   distance = col_double(),
##   sadmin = col_double(),
##   rulematch = col_double(),
##   lnfsize = col_double()
## )

ifri <- ifri_car_liv
str(ifri)
## Classes 'spec_tbl_df', 'tbl_df', 'tbl' and 'data.frame': 100 obs.
##     of  10 variables:
```

```
##   $ forest_id: num   217 325 88 174 240 287 324 321 216 82 ...
##   $ cid      : chr   "NEP" "IND" "UGA" "NEP" ...
##   $ zliv     : num   -0.614 -0.654 -0.338 -0.786 -0.45 ...
##   $ zbio     : num   -0.451 -0.365 -0.97 -1.325 -1.049 ...
##   $ livcar1  : num   3 3 3 3 3 3 3 3 3 3 ...
##   $ ownstate : num   1 1 1 1 1 1 1 1 1 0 ...
##   $ distance : num   2 1 1 2 2 1 2 2 2 1 ...
##   $ sadmin   : num   0 1 3 26 3 40 8 0 0 0 ...
##   $ rulematch: num   0 0 0 0 1 1 0 0 1 1 ...
##   $ lnfsize  : num   4.43 8.2 4.94 5.29 4.34 ...
##   - attr(*, "spec")=
##    .. cols(
##    ..     forest_id = col_double(),
##    ..     cid = col_character(),
##    ..     zliv = col_double(),
##    ..     zbio = col_double(),
##    ..     livcar1 = col_double(),
##    ..     ownstate = col_double(),
##    ..     distance = col_double(),
##    ..     sadmin = col_double(),
##    ..     rulematch = col_double(),
##    ..     lnfsize = col_double()
##    .. )
```

3.4 Data in a Stata File

The zip file for Chhatre and Agrawal's (2009) paper also contains a Stata dataset.

We go to the environment window and click on Import Dataset, then from Stata, or we run the following code to use the haven package (Wickham and Miller 2019):

```
library(haven)
ifri_car_liv <- read_dta("ifri_car_liv.dta")
```

We then proceed as before:

```
str(ifri)
## Classes 'spec_tbl_df', 'tbl_df', 'tbl' and 'data.frame':
##                        100 obs. of  10 variables:
##   $ forest_id: num   217 325 88 174 240 287 324 321 216 82 ...
##   $ cid      : chr   "NEP" "IND" "UGA" "NEP" ...
##   $ zliv     : num   -0.614 -0.654 -0.338 -0.786 -0.45 ...
##   $ zbio     : num   -0.451 -0.365 -0.97 -1.325 -1.049 ...
##   $ livcar1  : num   3 3 3 3 3 3 3 3 3 3 ...
##   $ ownstate : num   1 1 1 1 1 1 1 1 1 0 ...
##   $ distance : num   2 1 1 2 2 1 2 2 2 1 ...
##   $ sadmin   : num   0 1 3 26 3 40 8 0 0 0 ...
##   $ rulematch: num   0 0 0 0 1 1 0 0 1 1 ...
##   $ lnfsize  : num   4.43 8.2 4.94 5.29 4.34 ...
##   - attr(*, "spec")=
```

```
##     .. cols(
##     ..     forest_id = col_double(),
##     ..     cid = col_character(),
##     ..     zliv = col_double(),
##     ..     zbio = col_double(),
##     ..     livcar1 = col_double(),
##     ..     ownstate = col_double(),
##     ..     distance = col_double(),
##     ..     sadmin = col_double(),
##     ..     rulematch = col_double(),
##     ..     lnfsize = col_double()
##     .. )
```

3.5   Data from the World Development Indicators

The WDI package (Arel-Bundock 2019) will download the World Development Indicators of the World Bank. We use the `WDIsearch()` function to search for indicators. We should not have gaps in the keywords; note how we have used . and * below:

```
library(WDI)
WDIsearch("gdp.*capita.*PPP")
##        indicator
## [1,] "6.0.GDPpc_constant"
## [2,] "NY.GDP.PCAP.PP.KD.ZG"
## [3,] "NY.GDP.PCAP.PP.KD.87"
## [4,] "NY.GDP.PCAP.PP.KD"
## [5,] "NY.GDP.PCAP.PP.CD"
##        name
## [1,] "GDP per capita, PPP (constant 2011 international $) "
## [2,] "GDP per capita, PPP annual growth (%)"
## [3,] "GDP per capita, PPP (constant 1987 international $)"
## [4,] "GDP per capita, PPP (constant 2011 international $)"
## [5,] "GDP per capita, PPP (current international $)"
WDIsearch("CO2.*capita")
##        indicator
## [1,] "EN.ATM.CO2E.PC"
## [2,] "EN.ATM.NOXE.PC"
## [3,] "EN.ATM.METH.PC"
##        name
## [1,] "CO2 emissions (metric tons per capita)"
## [2,] "Nitrous oxide emissions (metric tons of CO2 equivalent
##        per capita)"
## [3,] "Methane emissions (kt of CO2 equivalent per capita)"
```

We are interested in the indicators "GDP per capita, PPP (constant 2011 international $)" and "CO2 emissions (metric tonnes per capita)". Once we have identified the specific indicators we would like to download, we use the `WDI()` function to download the data, and provide the indicator code.

```
wdi_data <- WDI(indicator =
    c("NY.GDP.PCAP.PP.KD",
      "EN.ATM.CO2E.PC"),
    start = 2010,
    end = 2010,
    extra = TRUE)
names(wdi_data)
##  [1] "iso2c"            "country"
##  [3] "year"             "NY.GDP.PCAP.PP.KD"
##  [5] "EN.ATM.CO2E.PC"   "iso3c"
##  [7] "region"           "capital"
##  [9] "longitude"        "latitude"
## [11] "income"           "lending"
```

☞ **Your Turn** Use `WDIsearch()` to search for an indicator of interest
to you, and then use `WDI()` to download the data.

3.6 Resources

Grolemund and Wickham (2017) have a chapter on importing data.

References

Arel-Bundock, V. 2019. WDI: World Development Indicators (World Bank). R Package Version
    2.6.0. https://CRAN.R-project.org/package=WDI.
Chhatre, A., and A. Agrawal. 2009. Trade-Offs and Synergies Between Carbon Storage and Liveli-
    hood Benefits from Forest Commons. *PNAS* 106 (42): 17667–17670.
Grolemund, G., and H. Wickham. 2017. *R for Data Science*. Boston: O'Reilly.
Shea, J.M. 2018. wooldridge: 111 Data Sets from "Introductory Econometrics: A Modern
    Approach, 6e" by Jeffrey M. Wooldridge. R Package Version 1.3.1. https://CRAN.R-project.
    org/package=wooldridge.
Wickham, H., J. Hester, and R. Francois. 2018. readr: Read Rectangular Text Data. R Package
    Version 1.3.1. https://CRAN.R-project.org/package=readr.
Wickham, H., and E. Miller. 2019. haven: Import and Export 'SPSS', 'Stata' and 'SAS' Files.
    R Package Version 2.1.1. https://CRAN.R-project.org/package=haven.

Wrangling and Graphing Data

4

4.1 Introduction

Data science is centred around data. A good way to know the characteristics of data is to graph our data. Tukey (1962, p. 49) had written: 'The simple graph has brought more information to the data analyst's mind than any other device'.

4.2 Example: Anscombe's Synthetic Data

Anscombe (1973) wrote a paper to show the value of graphing data, and this is available in R. We load the `tidyverse` package (Wickham 2017).

```
library(tidyverse)
```

The Anscombe data is available in R. We call it ans and use the `str` function to get a view of its structure.

```
data(anscombe)
ans <- anscombe
str(ans)
## 'data.frame':    11 obs. of  8 variables:
##  $ x1: num  10 8 13 9 11 14 6 4 12 7 ...
##  $ x2: num  10 8 13 9 11 14 6 4 12 7 ...
##  $ x3: num  10 8 13 9 11 14 6 4 12 7 ...
##  $ x4: num  8 8 8 8 8 8 8 19 8 8 ...
##  $ y1: num  8.04 6.95 7.58 8.81 8.33 ...
##  $ y2: num  9.14 8.14 8.74 8.77 9.26 8.1 6.13 3.1 9.13 7.26 ...
##  $ y3: num  7.46 6.77 12.74 7.11 7.81 ...
##  $ y4: num  6.58 5.76 7.71 8.84 8.47 7.04 5.25 12.5 5.56 7.91 ...
```

© Springer Nature Singapore Pte Ltd. 2020
V. Dayal, *Quantitative Economics with R*,
https://doi.org/10.1007/978-981-15-2035-8_4

We convert it to a tibble (the tidyverse version of a data frame) and have a glimpse:

```
ans <- as_tibble(anscombe)
glimpse(ans)
## Observations: 11
## Variables: 8
## $ x1 <dbl> 10, 8, 13, 9, 11, 14, 6, 4, 12, 7, 5
## $ x2 <dbl> 10, 8, 13, 9, 11, 14, 6, 4, 12, 7, 5
## $ x3 <dbl> 10, 8, 13, 9, 11, 14, 6, 4, 12, 7, 5
## $ x4 <dbl> 8, 8, 8, 8, 8, 8, 8, 19, 8, 8, 8
## $ y1 <dbl> 8.04, 6.95, 7.58, 8.81, 8.33, 9.96...
## $ y2 <dbl> 9.14, 8.14, 8.74, 8.77, 9.26, 8.10...
## $ y3 <dbl> 7.46, 6.77, 12.74, 7.11, 7.81, 8.8...
## $ y4 <dbl> 6.58, 5.76, 7.71, 8.84, 8.47, 7.04...
```

Since the tibble is small we can see it all.

```
ans
## # A tibble: 11 x 8
##        x1    x2    x3    x4    y1    y2    y3    y4
##     <dbl> <dbl> <dbl> <dbl> <dbl> <dbl> <dbl> <dbl>
## 1     10    10    10     8  8.04  9.14  7.46  6.58
## 2      8     8     8     8  6.95  8.14  6.77  5.76
## 3     13    13    13     8  7.58  8.74 12.7   7.71
## 4      9     9     9     8  8.81  8.77  7.11  8.84
## 5     11    11    11     8  8.33  9.26  7.81  8.47
## 6     14    14    14     8  9.96  8.1   8.84  7.04
## 7      6     6     6     8  7.24  6.13  6.08  5.25
## 8      4     4     4    19  4.26  3.1   5.39 12.5
## 9     12    12    12     8 10.8   9.13  8.15  5.56
## 10     7     7     7     8  4.82  7.26  6.42  7.91
## 11     5     5     5     8  5.68  4.74  5.73  6.89
```

We summarize the data, finding means.

```
ans %>%
  summarize(mean.x1 = mean(x1),
            mean.x2 = mean(x2),
            mean.y1 = mean(y1),
            mean.y2 = mean(y2))
## # A tibble:  1 x 4
##    mean.x1 mean.x2 mean.y1 mean.y2
##      <dbl> <dbl>     <dbl>   <dbl>
## 1        9     9   7.50 7.50
```

We see that the mean of x1 is the same as the mean of x2, and the mean of y1 is the same as the mean of y2.

We check the standard deviations of x1 and x2, y1 and y2.

```
ans %>%
  summarize(sd.x1 = sd(x1),
            sd.x2 = sd(x2),
            sd.y1 = sd(y1),
```

```
              sd.y2 = sd(y2))
## # A tibble: 1 x 4
##   sd.x1 sd.x2 sd.y1 sd.y2
##   <dbl> <dbl> <dbl> <dbl>
## 1  3.32  3.32  2.03  2.03
```

We see that the standard deviation of x1 is the same as the standard deviation of x2, and the standard deviation of y1 is the same as the standard deviation of y2.

We regress y1 on x1, and y2 on x2, using the lm() function.

```
mod1 <- lm(y1 ~ x1, data = ans)


mod2 <- lm(y2 ~ x2, data = ans)
```

Note that lm() is used to create objects, mod1 and mod2. We use the texreg package to tabulate the regression results.

```
library(texreg)
texreg(list(mod1, mod2),
       custom.model.names = c("mod1", "mod2"),
       caption = "Regressions of y1 on x1 and y2 on x2",
       caption.above = TRUE)
```

The regression coefficients, the R-squared statistic, etc. are the same in mod1 and mod2 in Table 4.1. We would expect the scatter plots of y1 versus x1 to be similar to the scatter plot of y2 versus x2. We now make a scatter plot of y1 versus x1 (Fig. 4.1).

```
ggplot(ans, aes(x = x1, y = y1)) +
  geom_point() +
  geom_smooth(method = "lm", se = FALSE)
```

We also make a scatter plot of y2 versus x2.

Table 4.1 Regressions of y1 on x1 and y2 on x2

	mod1	mod2
(Intercept)	3.00*	3.00*
	(1.12)	(1.13)
x1	0.50**	
	(0.12)	
x2		0.50**
		(0.12)
R^2	0.67	0.67
Adj. R^2	0.63	0.63
Num. obs.	11	11
RMSE	1.24	1.24

$^*p < 0.05, ^{**}p < 0.01, ^{***}p < 0.001$

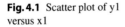

Fig. 4.1 Scatter plot of y1 versus x1

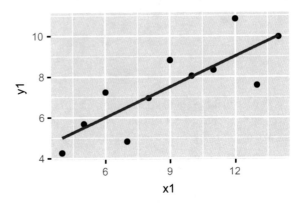

Fig. 4.2 Scatter plot of y2 versus x2

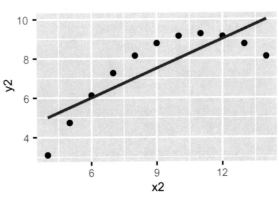

```
ggplot(ans, aes(x = x2, y = y2)) +
   geom_point()+
   geom_smooth(method = "lm", se = FALSE)
```

Figure 4.2 shows the scatter plot of y2 versus x2, and is very different from Fig. 4.1. There is a clear lack of fit in Fig. 4.2, because the true relationship is nonlinear and we have fitted a straight line.

☞ **Your Turn**   Check the means and standard deviations of x3, x4 and y3, y4. Also regress y3 on x3 and y4 on x4. Make scatter plots of y3 versus x3 and y4 versus x4. How do your results compare with those above?

4.3   Example: Carbon and Livelihoods Data

Forests are natural assets that provide multiple benefits—for example, carbon storage globally and local livelihood benefits (grazing and fuelwood, etc.) in developing countries. Chhatre and Agrawal (2009) analysed data relating to forests in developing countries. We saw how we can get their data into R in an earlier chapter.

```
ifri <- read_csv("ifri_car_liv.csv")
```

```
## Parsed with column specification:
## cols(
##   forest_id = col_double(),
##   cid = col_character(),
##   zliv = col_double(),
##   zbio = col_double(),
##   livcar1 = col_double(),
##   ownstate = col_double(),
##   distance = col_double(),
##   sadmin = col_double(),
##   rulematch = col_double(),
##   lnfsize = col_double()
## )
```

The data is available for 80 observations, rows 80 to 100 are blank. Below we print rows 79 to 81 of the data:

```
ifri[79:81,]
## # A tibble: 3 x 10
##    forest_id cid      zliv   zbio livcar1 ownstate
##        <dbl> <chr>   <dbl>  <dbl>   <dbl>    <dbl>
## 1        281 MAD     0.672 -0.184       4        1
## 2        169 NEP     1.22  -1.10        4        1
## 3         NA <NA>    NA     NA         NA       NA
## # ... with 4 more variables: distance <dbl>,
## #   sadmin <dbl>, rulematch <dbl>, lnfsize <dbl>
```

We keep only the first 80 rows.

```
# only 80 rows of data
ifri <- ifri[1:80,]
```

We rename the key variables so the output is easier to understand. The variable liveli is an index of forest livelihoods (fuelwood extraction, fodder, etc. from the forest). The variable carbon is a standardized measure of carbon storage, based on the basal area of trees per hectare.

```
ifri <- ifri %>%
  rename(carbon = zbio, liveli = zliv)
```

We plot a histogram and boxplot of the livelihood index variable (Figs. 4.3 and 4.4).

Fig. 4.3 Histogram of liveli

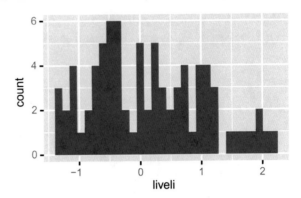

Fig. 4.4 Boxplot of liveli

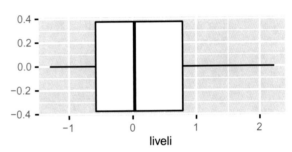

```
ggplot(ifri, aes(x = liveli)) +
  geom_histogram()
```

```
ggplot(ifri, aes(y = liveli)) +
  geom_boxplot() +
  coord_flip()
```

We plot a histogram and boxplot of the carbon index variable (Figs. 4.5 and 4.6).

```
ggplot(ifri, aes(x = carbon)) +
  geom_histogram()
```

```
## 'stat_bin()' using 'bins = 30'. Pick better
## value with 'binwidth'.
```

```
ggplot(ifri, aes(y = carbon)) +
  geom_boxplot() +
  coord_flip()
```

We make a scatter plot of carbon versus livelihoods (Fig. 4.7).

```
ggplot(ifri, aes(x = liveli,
      y = carbon)) +
  geom_point() +
  geom_smooth(method = "lm", se = FALSE)
```

Fig. 4.5 Histogram of carbon

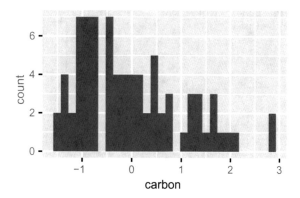

Fig. 4.6 Boxplot of carbon

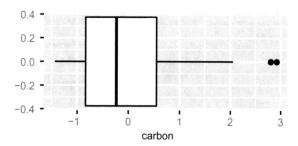

Fig. 4.7 Scatter plot of carbon versus livelihoods

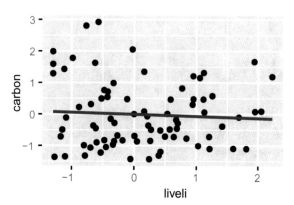

The relationship is weak in Fig. 4.7. A priori, we may have expected a negative relationship. The weak relationship suggests that we may not have a strong tradeoff between carbon and livelihoods; in some forests there may be scope to improve either or both with appropriate management.

We see how carbon and livelihoods are related to ownership.

```
ggplot(ifri, aes(x = factor(ownstate),
       y = carbon)) +
  geom_boxplot() + coord_flip()
```

Fig. 4.8 Boxplots of carbon
by ownership by state

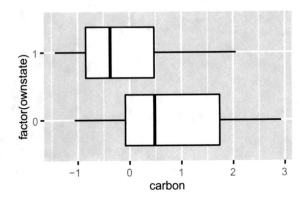

Fig. 4.9 Boxplots of
livelihood by ownership by
state

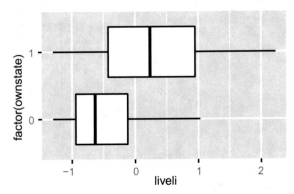

```
ggplot(ifri, aes(x = factor(ownstate),
      y = liveli)) +
  geom_boxplot() + coord_flip()
```

We see that state-owned forests have a lower distribution of carbon in the sample
(Fig. 4.8). We see that state-owned forests have a higher distribution of livelihoods
in the sample (Fig. 4.9).

```
ggplot(ifri, aes(x = factor(rulematch),
      y = carbon)) +
  geom_boxplot() + coord_flip()
```

The distribution of carbon is higher when rule match is 1. When rule match is 1
(Figs. 4.10 and 4.11), there is a better perception of the rules.

```
ggplot(ifri, aes(x = factor(rulematch),
      y = liveli)) +
  geom_boxplot() + coord_flip()
```

Fig. 4.10 Boxplots of carbon by rule match

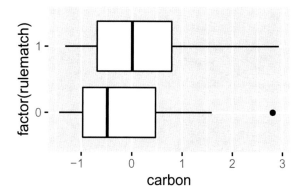

Fig. 4.11 Boxplots of livelihood by rule match

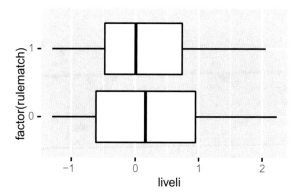

We now create a variable that combines ownership (state or community) and rule match.

```
ifri2 <- mutate(ifri, f_own_rule =
  ifelse(ownstate == 1 & rulematch == 0, "State_low",
  ifelse(ownstate == 1 & rulematch == 1, "State_high",
  ifelse(ownstate == 0 & rulematch == 1, "Com_high",
  "Com_low"        ))))

ggplot(ifri2, aes(x = liveli,
            y = carbon,
            size = lnfsize,
            colour = f_own_rule)) +
  geom_point() +
  geom_smooth(method = "lm",
            se=FALSE)
```

Figure 4.12 illustrates the power of ggplot. We can see the relationship between carbon and livelihoods while incorporating (1) forest size (represented by the size of the bubbles), (2) ownership and (3) rule match in the plot.

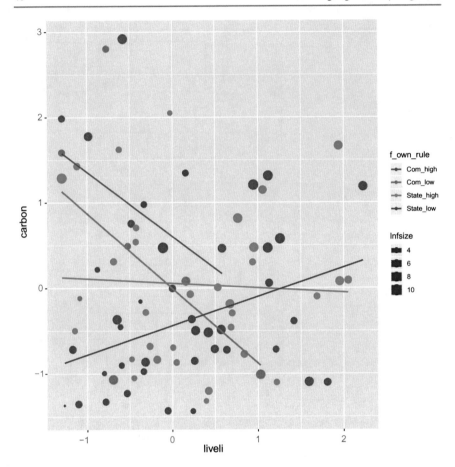

Fig. 4.12 Scatter plot of carbon versus livelihood

☞ **Your Turn**   Look at Fig. 4.12 carefully, writing down different aspects
that strike you.

4.4   Example: WDI Data on CO2 and Per Capita Income

4.4.1   Getting the Data

We had examined how the WDI package (Arel-Bundock 2019) can be used to load
data into R. We first search for indicators related to per capita GDP and CO2.

```
library(WDI)
#new_wdi_cache <- WDIcache()
WDIsearch("gdp.*capita.*PPP")
##       indicator
## [1,] "6.0.GDPpc_constant"
## [2,] "NY.GDP.PCAP.PP.KD.ZG"
## [3,] "NY.GDP.PCAP.PP.KD.87"
## [4,] "NY.GDP.PCAP.PP.KD"
## [5,] "NY.GDP.PCAP.PP.CD"
##       name
## [1,] "GDP per capita, PPP (constant 2011 international $) "
## [2,] "GDP per capita, PPP annual growth (%)"
## [3,] "GDP per capita, PPP (constant 1987 international $)"
## [4,] "GDP per capita, PPP (constant 2011 international $)"
## [5,] "GDP per capita, PPP (current international $)"
#,cache = new_wdi_cache)
WDIsearch("CO2.*capita")
##       indicator
## [1,] "EN.ATM.CO2E.PC"
## [2,] "EN.ATM.NOXE.PC"
## [3,] "EN.ATM.METH.PC"
##       name
## [1,] "CO2 emissions (metric tons per capita)"
## [2,] "Nitrous oxide emissions (metric tons of CO2 equivalent per
##        capita)"
## [3,] "Methane emissions (kt of CO2 equivalent per capita)"
#,  cache = new_wdi_cache)
```

We choose our indicators:

```
wdi_data <- WDI(indicator =
    c("NY.GDP.PCAP.PP.KD",
      "EN.ATM.CO2E.PC"),
    start = 2010,
    end = 2010,
    extra = TRUE)
```

Having loaded the data into R, we wrangle it. Countries are the units of analysis. Since we downloaded data for all countries, we need to remove aggregates, so that we only have data relating to countries.

```
library(tidyverse)

wdi_data <- wdi_data %>%
  filter(region != "Aggregates")
```

We rename our variables.

```
wdi_data <- wdi_data %>%
rename(GDPpercap =
         NY.GDP.PCAP.PP.KD,
       Emit_CO2percap =
         EN.ATM.CO2E.PC)
```

We write our data as a csv file, so that we have it on our computer.

```
write_csv(wdi_data,"wdi_CO2_GDP.csv")
```

We read the csv file into R.

```
wdi <- read_csv("wdi_CO2_GDP.csv")
```

```
## Parsed with column specification:
## cols(
##   iso2c = col_character(),
##   country = col_character(),
##   year = col_double(),
##   GDPpercap = col_double(),
##   Emit_CO2percap = col_double(),
##   iso3c = col_character(),
##   region = col_character(),
##   capital = col_character(),
##   longitude = col_double(),
##   latitude = col_double(),
##   income = col_character(),
##   lending = col_character()
## )
```

4.4.2   Graphing the Data

We ask for a summary of GDP per capita.

```
summary(wdi$GDPpercap)
##      Min.  1st Qu.   Median     Mean  3rd Qu.
##     659.8   3744.5  10176.8  18075.0  27910.8
##      Max.     NA's
## 119973.6       22
```

The mean is higher than the median—the data is positively skewed.

```
ggplot(wdi,
       aes(x = GDPpercap)) +
  geom_histogram()
```

```
## 'stat_bin()' using 'bins = 30'. Pick better
## value with 'binwidth'.
## Warning: Removed 22 rows containing non-finite values
## (stat_bin).
```

Fig. 4.13 Histogram of GDP
per capita

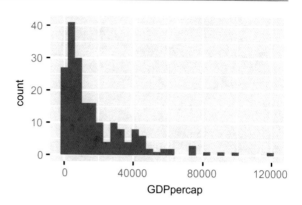

The histogram (Fig. 4.13) shows that there are a few countries that are very
wealthy. Figure 4.14 shows the region-wise distribution of GDP per capita. Sub-
Saharan Africa and South Asia are the world's poorest regions.

```
ggplot(wdi,
       aes(y = GDPpercap,
       x = region)) +
  geom_boxplot() +
  coord_flip() +
  scale_y_log10()
```

```
## Warning: Removed 22 rows containing non-finite values
## (stat_boxplot).
```

```
summary(wdi$Emit_CO2percap)
##      Min.  1st Qu.   Median     Mean  3rd Qu.
##   0.02452  0.65234  2.59865  4.94310  6.64236
##      Max.     NA's
## 39.05971       14
```

☞ **Your Turn**   Make a histogram of CO2percap and a boxplot of
CO2percap by region.

We make a scatter plot of GDP per capita versus CO2 emissions per capita.

```
gg1 <- ggplot(wdi,
        aes(x=GDPpercap,
            y=Emit_CO2percap)) +
  geom_point()
gg1
```

```
## Warning: Removed 28 rows containing missing values
## (geom_point).
```

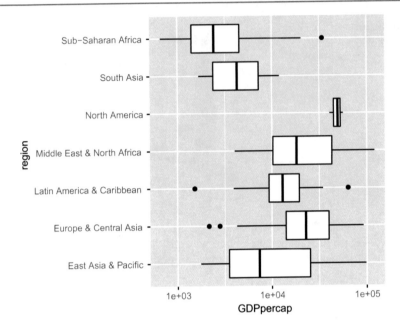

Fig. 4.14 Boxplot of GDP per capita by region

Fig. 4.15 Scatter plot of
GDP per capita versus CO2
emissions per capita

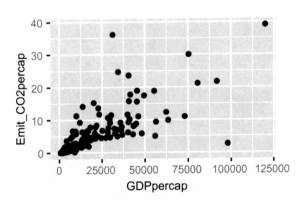

The data are clustered in the bottom left corner and very spread out away from
there (Fig. 4.15). To get a more even distribution of the data, we use axes with log
scales. We also add a smooth:

```
gg2 <- gg1 + geom_smooth(se = FALSE) +
  scale_x_log10() +
  scale_y_log10()

gg2

## 'geom_smooth()' using method = 'loess' and formula 'y ~ x'
```

Fig. 4.16 Scatter plot of
GDP per capita versus CO2
emissions per capita

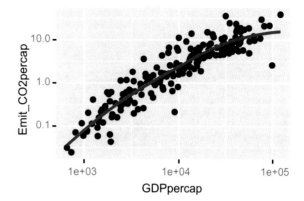

```
## Warning: Removed 28 rows containing non-finite values
## (stat_smooth).
## Warning: Removed 28 rows containing missing values
## (geom_point).
```

Richer countries emit more CO2 per capita, and the association is strong. There is some reduction in the slope of the smooth curve as GDP per capita increases (Fig. 4.16).

4.4.3 Mapping the Data

Quite remarkably, we can draw maps with the ggplot2 package, with some help from the maps package (Becker et al. 2018). The maps package has some basic spatial data in it.

```
#install.packages("maps")
library(maps)
```

We ask for map data related to the world.

```
dat_map <- map_data("world")
dim(dat_map)
## [1] 99338      6
class(dat_map)
## [1] "data.frame"
head(dat_map)
##          long      lat group order  region subregion
## 1 -69.89912 12.45200     1     1   Aruba      <NA>
## 2 -69.89571 12.42300     1     2   Aruba      <NA>
## 3 -69.94219 12.43853     1     3   Aruba      <NA>
## 4 -70.00415 12.50049     1     4   Aruba      <NA>
## 5 -70.06612 12.54697     1     5   Aruba      <NA>
## 6 -70.05088 12.59707     1     6   Aruba      <NA>
```

We can make a blank world map using `geom_polygon`.

```
ggplot(dat_map, aes(x = long, y = lat,
                    group = group)) +
  geom_polygon(fill = "white", colour = "black")
```

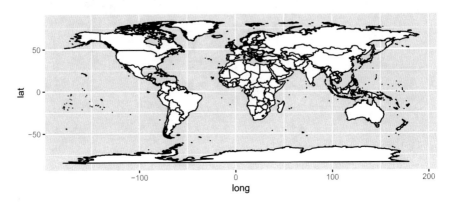

We have WDI data and the data in the maps package. We need to translate the way country names are recorded in both sets of data; for which we use the countrycode package (Arel-Bundock et al. 2018).

```
#install.packages("countrycode")

library(countrycode)


dat_map$ccode <- countrycode(dat_map$region,
       origin = "country.name",
       destination = "wb")


## Warning in countrycode(dat_map$region, origin = "country.
name", destination = "wb"): Some values were not matched
unambiguously: Anguilla, Antarctica, Ascension Island, Azores,
Barbuda, Bonaire, Canary Islands, Chagos Archipelago, Christmas
Island, Cocos Islands, Cook Islands, Falkland Islands, French
Guiana, French Southern and Antarctic Lands, Grenadines,
Guadeloupe, Guernsey, Heard Island, Jersey, Madeira Islands,
Martinique, Mayotte, Micronesia, Montserrat, Niue, Norfolk
Island, Pitcairn Islands, Reunion, Saba, Saint Barthelemy,
Saint Helena, Saint Martin, Saint Pierre and Miquelon,
Siachen Glacier, Sint Eustatius, Somalia, South Georgia,
South Sandwich Islands, Vatican, Virgin Islands, Wallis and
Futuna, Western Sahara
```

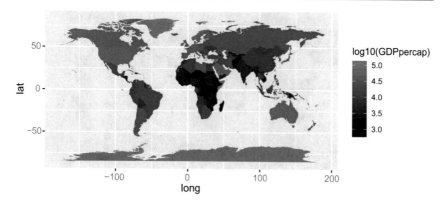

Fig. 4.17 Global distribution of GDP per capita

```
wdi$ccode <- countrycode(wdi$country,
       origin = "country.name",
       destination = "wb")
```

```
## Warning in countrycode(wdi$country, origin = "country.name",
destination = "wb"): Some values were not matched unambiguously:
Eswatini, Somalia
```

Now we merge the two sets of data:

```
merged <- full_join(dat_map, wdi,
                    by = "ccode")
```

We make a map of the global distribution of GDP per capita (Fig. 4.17).

```
ggplot(merged, aes(x = long, y = lat,
      group = group, fill = log10(GDPpercap))) +
  geom_polygon()
```

We use a colour gradient (Fig. 4.18).

```
ggplot(merged, aes(x = long, y = lat,
      group = group,
      fill = log10(GDPpercap))) +
  geom_polygon() +
  scale_fill_gradient(low = "green",
                      high = "red")
```

We also make a similar map for CO2 (Fig. 4.19).

```
ggplot(merged, aes(x = long, y = lat,
      group = group,
      fill = log10(Emit_CO2percap))) +
  geom_polygon() +
  scale_fill_gradient(low = "green",
                      high = "red")
```

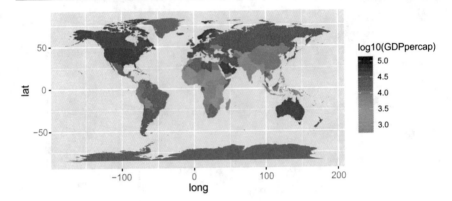

Fig. 4.18 Global distribution of GDP per capita

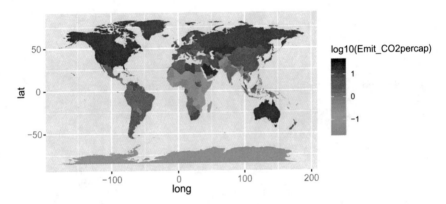

Fig. 4.19 Global distribution of CO2 emissions per capita

4.5   Resources

For Better Understanding

The online Datacamp course (https://www.datacamp.com/courses/introduction-to-the-tidyverse), Introduction to the Tidyverse, by David Robinson, is very useful if you are new to the tidyverse. Robinson shows the value of wrangling and graphing in tandem.

For Going Further

Grolemund and Wickham (2017) is *the* book to develop skills in wrangling and graphing data (Table 4.2).

Healy (2019) book is a detailed practical introduction to data visualization, covering both why and how, and uses the `ggplot2` package.

Table 4.2 Regressions of y3 on x3 and y4 on x4

	mod3	mod4
(Intercept)	3.00*	3.00*
	(1.12)	(1.12)
x3	0.50**	
	(0.12)	
x4		0.50**
		(0.12)
R^2	0.67	0.67
Adj. R^2	0.63	0.63
Num. obs.	11	11
RMSE	1.24	1.24

$^*p < 0.05$, $^{**}p < 0.01$, $^{***}p < 0.001$

Answers to Your Turn

☞ **Your Turn** (Sect. 4.2)   Check the means and standard deviations of x3, x4 and y3, y4. Also regress y3 on x3 and y4 on x4. Make scatter plots of y3 versus x3 and y4 versus x4. How do your results compare with those above?

```
data(anscombe)
ans <- anscombe


ans %>%
  summarize(mean.x3 = mean(x3),
            mean.x4 = mean(x4),
            mean.y3 = mean(y3),
            mean.y4 = mean(y4))
##    mean.x3 mean.x4 mean.y3  mean.y4
## 1       9       9     7.5 7.500909


mod3 <- lm(y3 ~ x3, data = ans)


mod4 <- lm(y4 ~ x4, data = ans)


library(texreg)
texreg(list(mod3, mod4),
       custom.model.names = c("mod3", "mod4"),
       caption = "Regressions of y3 on x3 and y4 on x4",
       caption.above = TRUE)
```

See Table 4.2.

```
ggplot(ans, aes(x = x3, y = y3)) +
  geom_point() +
  geom_smooth(method = "lm", se = FALSE)
```

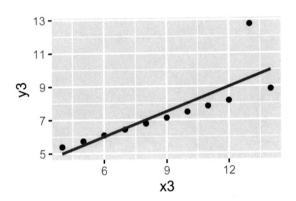

```
ggplot(ans, aes(x = x4, y = y4)) +
  geom_point() +
  geom_smooth(method = "lm", se = FALSE)
```

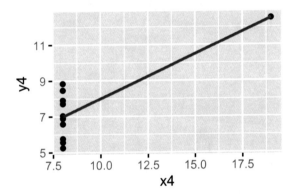

☞ **Your Turn** (Sect. 4.3)   Look at the relationship of carbon and livelihoods with forest size. Make scatter plots of
carbon and livelihoods versus lnfsize (which is
log of forest size).

```
ggplot(ifri, aes(x = lnfsize,
     y = carbon)) +
  geom_point() +
  geom_smooth(method = "lm", se = FALSE)
```

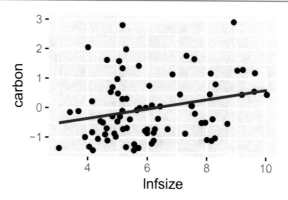

```
ggplot(ifri, aes(x = lnfsize,
    y = liveli)) +
  geom_point() +
  geom_smooth(method = "lm", se = FALSE)
```

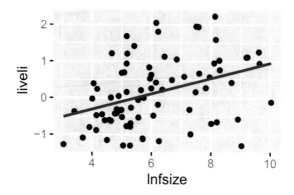

☞ **Your Turn** (Sect. 4.4)   Make a histogram of CO2percap and a boxplot of CO2percap by region.

```
ggplot(wdi,
    aes(x = Emit_CO2percap)) +
  geom_histogram()
```

```
## 'stat_bin()' using 'bins = 30'. Pick better
## value with 'binwidth'.
## Warning: Removed 14 rows containing non-finite values
## (stat_bin).
```

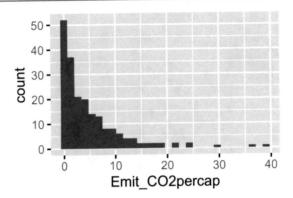

```
ggplot(wdi,
       aes(y = Emit_CO2percap,
           x = region)) +
  geom_boxplot() +
  coord_flip() +
  scale_y_log10()

## Warning: Removed 14 rows containing non-finite values
## (stat_boxplot).
```

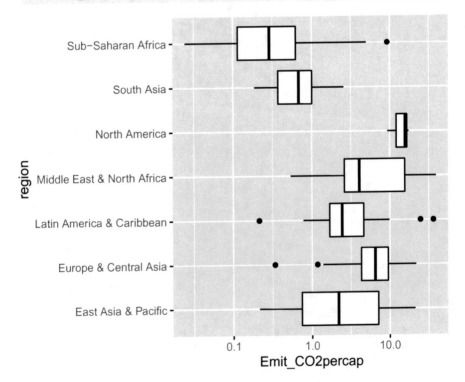

References

Anscombe, F.J. 1973. Graphs in statistical analysis. *The American Statistician* 27 (1): 17–21.

Arel-Bundock, V. 2019. WDI: World development indicators (World Bank). *R Package Version* 2 (6). https://CRAN.R-project.org/package=WDI.

Arel-Bundock, et al. 2018. Countrycode: An R package to convert country names and country codes. *Journal of Open Source Software* 3 (28): 848. https://doi.org/10.21105/joss.00848.

Becker, R.A., A.R. Wilks, R version by Ray Brownrigg. Enhancements by Thomas P Minka and Alex Deckmyn. 2018. maps: Draw Geographical Maps. *R Package Version* 3 (3). https://CRAN.R-project.org/package=maps.

Chhatre, A., A. Agrawal. 2009. Trade-offs and synergies between carbon storage and livelihood benefits from forest commons. *PNAS* 106 (42): 17667–17670.

Grolemund, G., H. Wickham. 2017. *R for Data Science*. Boston: O'Reilly.

Healy, K. 2019. *Data Visualization: A Practical Introduction*. Princeton: Princeton University Press.

Tukey, J. 1962. The future of data analysis. *The Annals of Mathematical Statistics*. 33: 1–67.

Wickham, H. 2017. tidyverse: Easily install and load the 'Tidyverse'. *R Package Version* 1 (2): 1. https://CRAN.R-project.org/package=tidyverse.

Networks

5

5.1 Introduction

One aspect of data science is the diversity of types of data. Usually the data we examine has measurements related to a unit of analysis—height and weight of a person, for example. In network data we emphasize the connections between units of analysis—do Deng and Anne talk to each other?

Reflecting on the growing importance of the study of networks in economics, Jackson (2014, pp. 3–4) writes:

> ... as economists endeavour to build better models of human behaviour, they cannot ignore that humans are fundamentally a social species with interaction patterns that shape their behaviors. ... Ultimately, the full network of relationships—how dense it is, whether some groups are segregated, who sits in central positions—affects how information spreads and how people behave.

R now has a rich set of packages for network analysis. The igraph package (Csardi and Nepusz 2006) is one of the leading packages. We will be using the tidygraph (Pedersen 2019) and ggnetwork (Briatte 2016) packages here; both have a tidyverse (Wickham 2017) approach to network analysis. The tidygraph package builds on the igraph (Csardi and Nepusz 2006) package. The intergraph package (Bojanowski 2015) is useful in converting network objects that work with different packages in R. For now, we load the tidyverse and tidygraph packages.

```
library(tidyverse)
library(tidygraph)
```

© Springer Nature Singapore Pte Ltd. 2020
V. Dayal, *Quantitative Economics with R*,
https://doi.org/10.1007/978-981-15-2035-8_5

5.2   Simple Example with Synthetic Data

We start with a simple hypothetical example. We have a set of people who talk to each other. Asif and Gita talk to each other—the graph is undirected—Asif is the first person in the `from` vector and Gita is the first person in the `to` vector.

```
from <- c("Asif","Deng","Gita","Paul","Sure",
          "Sure")
to <- c("Gita","Gita","Anne","Anne","Paul",
        "Asif")
```

We will be building on the `tidyverse` approach presented earlier in the book, adding network elements.

We create a tibble called edge1.

```
edge1 <- tibble(from,to)
str(edge1)
## Classes 'tbl_df', 'tbl' and 'data.frame':      6 obs. of  2
    variables:
##  $ from: chr  "Asif" "Deng" "Gita" "Paul" ...
##  $ to  : chr  "Gita" "Gita" "Anne" "Anne" ...
class(edge1)
## [1] "tbl_df"      "tbl"         "data.frame"
edge1
## # A tibble: 6 x 2
##    from  to
##    <chr> <chr>
## 1 Asif  Gita
## 2 Deng  Gita
## 3 Gita  Anne
## 4 Paul  Anne
## 5 Sure  Paul
## 6 Sure  Asif
```

The tibble `edge1` is a usual tibble, i.e. a kind of data frame.

The different packages for networks in R use different objects. We will conduct a series of operations starting with the object `edge1`. Before we move into the specific code, it may be useful to provide an overview:

1. `edge1` is a tibble/dataframe. We use the `tbl_graph` function to convert this to:
2. `Talk`, which is a `tbl_graph`. We use the `asNetwork` function to convert this to:
3. `Talk_n`, which is a `network`. We use the `ggnetwork` function to convert this to:
4. `Talk_g`, which is a dataframe. Note that this dataframe has information that will allow plotting. We use the function `ggplot` to convert this to:
5. `Talk_GG`, which is a `ggplot`, and gives us a plot of the network graph.

Getting back to `edge1`, we use the `tbl_graph` function of the `tidygraph` package to convert it into an object that stores information about nodes (here Asif, Deng etc.) and the edges (whether one person talks to another):

```
Talk <- tbl_graph(edges = edge1,
                  directed = FALSE)
Talk
## # A tbl_graph: 6 nodes and 6 edges
## #
## # An undirected simple graph with 1 component
## #
## # Node Data: 6 x 1 (active)
##   name
##   <chr>
## 1 Asif
## 2 Gita
## 3 Deng
## 4 Anne
## 5 Paul
## 6 Sure
## #
## # Edge Data: 6 x 2
##    from    to
##   <int> <int>
## 1     1     2
## 2     2     3
## 3     2     4
## # ... with 3 more rows
class(Talk)
## [1] "tbl_graph" "igraph"
```

To clarify: `tbl_graph` converted the tibble `edge1` to the `Talk` object, which is a `tbl_graph`, and which has 6 nodes (Asif, Gita etc.) and 6 edges (connections between the nodes).

We use `activate` to bring either the node or the edge data into play. To bring the edges into play:

```
Talk %>%
  activate(edges) %>%
  as_tibble()
## # A tibble: 6 x 2
##    from    to
##   <int> <int>
## 1     1     2
## 2     2     3
## 3     2     4
## 4     4     5
## 5     5     6
## 6     1     6
class(Talk)
## [1] "tbl_graph" "igraph"
```

Compare the output above with that below, where nodes are activated:

```
Talk %>%
  activate(nodes) %>%
  as_tibble()
## # A tibble: 6 x 1
##    name
##    <chr>
## 1 Asif
## 2 Gita
## 3 Deng
## 4 Anne
## 5 Paul
## 6 Sure
class(Talk)
## [1] "tbl_graph" "igraph"
```

We will now use the ggnetwork package to plot our little network. We need to first convert the Talk object to a network object that ggnetwork can work with. The intergraph package helps us convert objects, its asNetwork function changes Talk.

```
library(intergraph)
Talk_n <- asNetwork(Talk)
Talk_n
##  Network attributes:
##    vertices = 6
##    directed = FALSE
##    hyper = FALSE
##    loops = FALSE
##    multiple = FALSE
##    bipartite = FALSE
##    total edges= 6
##      missing edges= 0
##      non-missing edges= 6
##
##  Vertex attribute names:
##      vertex.names
##
## No edge attributes
class(Talk_n)
## [1] "network"
```

Now ggnetwork will convert Talk_n to a dataframe, which has information which will facilitate plotting by ggplot:

```
library(ggnetwork)
Talk_g <- ggnetwork(Talk_n)
class(Talk_g)
## [1] "data.frame"
```

We see that the ggnetwork function has created a dataframe that can be used for plotting the network using the ggplot2 package. Some new geoms—geom_edges, geom_nodes, and geom_nodetext—help us plot edges, nodes and text about nodes.

```
Talk_GG <- ggplot(Talk_g, aes(x,y, xend = xend,
                  yend = yend)) +
  geom_edges(color = "lightgrey") +
  geom_nodes(alpha = 0.6, size = 5) +
  geom_nodetext(aes(label = vertex.names),
                col = "blue") +
  theme_blank()
Talk_GG

class(Talk_GG)
## [1] "gg"      "ggplot"
```

Our initial graph of the network (Fig. 5.1) is not very clear. It can take some effort to get the visualisation of networks right. Depending on the context, and the size of the graph, we can use different geoms to get a plot that illuminates. We create our own function that we can use repeatedly, and which gives us a clearer plot, although it is suitable for networks that are not large.

```
ggnetplot <- function(Net = Bali) {
  Net <- ggnetwork(Net, layout =
                   "kamadakawai")
  ggplot(Net, aes(x,y, xend = xend,
                  yend = yend)) +
    geom_edges(col = "tomato") +
    # node text repelled from node
    geom_nodetext_repel(aes(label = vertex.names),
            col = "black", size = 3) +
    theme_blank()
}

ggnetplot(Talk_n)
```

Figure 5.2 is an improved version of Fig. 5.1.

☞ **Your Turn**   Create and plot a small network, where Tim talks to Jim,
Kim talks to Jim and Jane, and Jane talks to Jim. You can
start with:

Fig. 5.1 Initial graph of talk
network

Fig. 5.2 Improved diagram
for the talk network

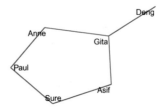

```
# Hint
from <-  c("Kim", "Tim", "Kim", "Jim")
to   <-  c("Jim", "Jim", "Jane", "Jane")
```

5.3   Example: Medici Network

We now graph a network of marriages among prominent families in Florence, that
was researched by Padgett and Ansell (1993, p. 1260), who analyzed the "early
15th-century rise of Cosimo de' Medici in Renaissance Florence." Padgett and Ansell
(1993, p. 1260) focused on "the structure and the sequential emergence of the marriage, economic, and patronage networks that constituted the Medicean political
party, used by Cosimo in 1434 to take over the budding Florentine Renaissance
state." The data are in the `netrankr` package (Schoch 2017).

```
library(netrankr)
data("florentine_m")
class(florentine_m)
## [1] "igraph"
```

We convert `florentine_m` to `flor` and then to `flor_g`:

```
flor <- as_tbl_graph(florentine_m)
class(flor)
## [1] "tbl_graph" "igraph"
flor_g <- asNetwork(flor)
class(flor_g)
## [1] "network"
```

We now plot the Florentine marriage network, using the function `ggnetplot`
that we created in the previous section (Fig. 5.3). Different families are displayed
(nodes), along with their relationships via marriages.

```
# ggnetplot uses ggnetwork() and ggplot()
ggnetplot(flor_g)
```

We now calculate a centrality index, using a simple if crude measure, degree,
with the `centrality_degree` function. Degree counts the number of edges that
come out of a given node. The Ginori family has one node coming out of it, so has
degree one (Fig. 5.3). We will use this measure through the chapter. Our code is on
the lines of `tidyverse` code.

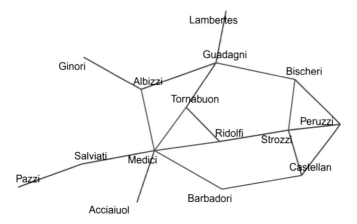

Fig. 5.3 Florentine marriage network

```
flor2 <- flor %>%
  activate(nodes) %>%
  mutate(degree = centrality_degree())
```

We extract the wealth and degree data by family and arrange the data by degree:

```
flor3 <- as_tibble(flor2) %>%
  arrange(-degree)
class(flor3)
## [1] "tbl_df"     "tbl"         "data.frame"
flor3
## # A tibble: 16 x 3
##     name       wealth degree
##     <chr>       <int>  <dbl>
##  1 Medici        103      6
##  2 Guadagni        8      4
##  3 Strozzi       146      4
##  4 Albizzi        36      3
##  5 Bischeri       44      3
##  6 Castellan      20      3
##  7 Peruzzi        49      3
##  8 Ridolfi        27      3
##  9 Tornabuon      48      3
## 10 Barbadori      55      2
## 11 Salviati       10      2
## 12 Acciaiuol      10      1
## 13 Ginori         32      1
## 14 Lambertes      42      1
## 15 Pazzi          48      1
## 16 Pucci           3      0
```

We can plot the relation between wealth and degree (Fig. 5.4) using the `flor3` object we created above. We use the `ggrepel` package (Slowikowski 2019) to get nice, uncluttered labels.

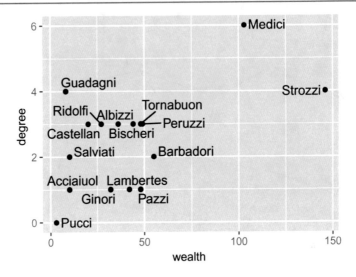

Fig. 5.4 Degree versus wealth of different families in Florence. Note the high wealth and degree of the Medici family

```
library(ggrepel)
ggplot(flor3, aes(x= wealth, y = degree,
                  label = name)) +
  geom_point() +
  # position labels appropriately
  geom_text_repel()
```

We see that the Medici family had high degree and wealth (Fig. 5.4). The Medici family had less wealth than the Strozzi, but had a higher degree.

5.4   Example: Bali Terrorist Network

We now plot another network, the Bali terrorist network. Koschade (2006) carried out a social network analysis of the Jemaah Islamiyah (an armed militant organization) cell that was responsible for the Bali bombings in 2002, in which 202 people were killed. Terrorist cells work in secrecy and who talks to whom is a key feature of how they work. Such an analysis helps understand their functioning.

The Bali data is a network object, available in the UserNetR (Luke 2018) package.

```
library(UserNetR)
data("Bali")
class(Bali)
## [1] "network"
```

We convert the network to a `tbl_graph`:

```
Bali_t <- as_tbl_graph(Bali)
Bali_t
## # A tbl_graph: 17 nodes and 63 edges
## #
## # An undirected simple graph with 1 component
## #
## # Node Data: 17 x 3 (active)
##    na    role  name
##    <lgl> <chr> <chr>
## 1 FALSE CT    Muklas
## 2 FALSE OA    Amrozi
## 3 FALSE OA    Imron
## 4 FALSE CT    Samudra
## 5 FALSE BM    Dulmatin
## 6 FALSE CT    Idris
## # ... with 11 more rows
## #
## # Edge Data: 63 x 4
##    from    to   IC na
##    <int> <int> <dbl> <lgl>
## 1     1     2     2 FALSE
## 2     1     3     2 FALSE
## 3     1     4     1 FALSE
## # ... with 60 more rows
```

There are 17 nodes and 63 edges. We plot the network with the function
ggnetplot we have created.

```
ggnetplot(Bali)
```

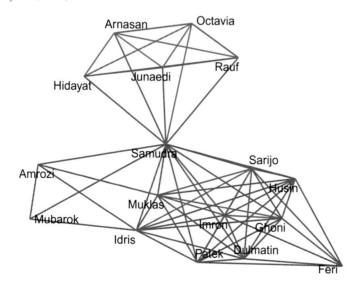

Fig. 5.5 Bali terrorist network

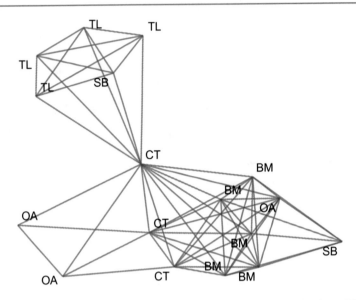

Fig. 5.6 Bali terrorist network, with roles, CT command team, OA operational assistant, BM, bomb maker, SB suicide bomber, TL team lima

Samudra (the field commander) was at the centre of the network (Fig. 5.5), with the greatest interaction with others. We now plot the network with nodes classified by the role of the member (Fig. 5.6); this gives us greater general insight compared to Fig. 5.5. Team Lima was the support group of the operation.

```
Bali_g <- ggnetwork(Bali)
ggplot(Bali_g, aes(x,y, xend = xend,
                   yend = yend)) +
    geom_edges(col = "tomato",
               alpha = 0.6) +
    geom_nodetext_repel(aes(label = role), size = 3) +
    theme_blank()
```

5.5   Simulating Network Formation

How do networks form? Two different models of network formation, which result in very different types of networks, are the (1) Erdos Renyi model of random networks where two nodes connect or do not with a given probability, at random, and (2) Barabasi and Albert model where networks form over time and each new node prefers to attach itself to well connected nodes.

We now plot some simulations of networks. We first consider an Erdos Renyi simulation, where the number of nodes is n and the probability of an edge connecting any two vertices is p. According to Barabasi and Bonabeau (2003, p. 62), "In 1959, aiming to describe networks seen in communications and the life sciences, Erdos

Fig. 5.7 Degree distribution
for Erdos Renyi simulation
with n = 30, p = 0.2

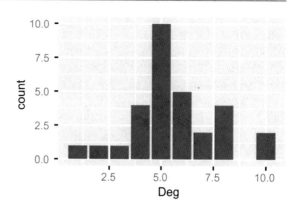

and Renyi suggested that such systems could be effectively modeled by connecting
their nodes with randomly placed links."

```
rg <- play_erdos_renyi(n = 30, p = 0.2,
                       directed = FALSE)
class(rg)
## [1] "tbl_graph" "igraph"
rg2 <- rg %>%
  activate(nodes) %>%
  mutate(Deg = centrality_degree())

rg3 <- rg2 %>%
  activate(nodes) %>%
  as_tibble
```

We plot the degree distribution emerging from the Erdos Renyi simulation
(Fig. 5.7). The distribution is bell shaped.

```
ggplot(rg3, aes(x = Deg)) +
  geom_bar()
```

The code below plots the network (Fig. 5.8).

```
rg_g <- asNetwork(rg2)
rg_g <- ggnetwork(rg_g, layout =
                  "kamadakawai")
ggplot(rg_g, aes(x,y, xend = xend,
                 yend = yend)) +
  geom_edges(col = "tomato") +
  geom_nodes(aes(size = Deg),
             alpha = 0.4, #size = 1,
             col = "black") +
  theme_blank()
```

☞ **Your Turn**   Use the code above to conduct an Erdos Renyi simulation
with n = 500 and p = 0.05. What do you observe?

Fig. 5.8 Network diagram
for Erdos Renyi simulation
with n = 10 and p = 0.2

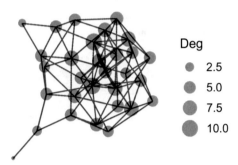

Deg

● 2.5

● 5.0

● 7.5

● 10.0

Barabasi and Albert (1999) proposed a model based on the concept of preferential attachment. A new node joining a network will prefer to attach itself to nodes that are more connected. They wrote (1999, p. 509), "Systems as diverse as genetic networks or the World Wide Web are best described as networks with complex topology. A common property of many large networks is that the vertex connectivities follow a scale-free power-law distribution. This feature was found to be a consequence of two generic mechanisms: (i) networks expand continuously by the addition of new vertices, and (ii) new vertices attach preferentially to sites that are already well connected."

We now plot graphs relating to simulation of such a model of network formation.

```
ba <- play_barabasi_albert(n = 30, power = 1,
              directed = FALSE)
class(ba)
## [1] "tbl_graph" "igraph"
ba2 <- ba %>%
  activate(nodes) %>%
  mutate(Deg = centrality_degree())


ba3 <- ba2 %>%
  activate(nodes) %>%
  as_tibble
```

```
ggrg3 <- ggplot(rg3, aes(x = Deg)) +
  geom_bar()
ggba3 <- ggplot(ba3, aes(x = Deg)) +
  geom_bar()
library(gridExtra)
grid.arrange(ggrg3, ggba3, ncol = 2)
```

Figure 5.9 contrasts the degree distribution of the networks arising out of the Erdos Renyi and Barabasi Albert models.

```
ba_g <- asNetwork(ba2)

  ba_g <- ggnetwork(ba_g, layout =
            "kamadakawai")
```

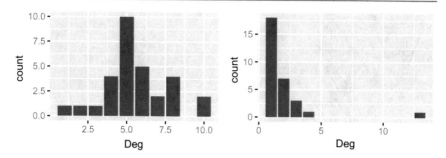

Fig. 5.9 Degree distribution for Erdos Renyi (left) and Barabasi Albert (right). In the former, nodes are connected by edges randomly, and with a given probability; in the latter as the network grows, a new node is more likely to attach itself to a node that is well connected

```
ggrg <- ggplot(rg_g, aes(x,y, xend = xend,
                  yend = yend)) +
    geom_edges(col = "tomato") +
    geom_nodes(aes(size = Deg),
              alpha = 0.4, #size = 1,
              col = "black") +
    theme_blank()

ggba <- ggplot(ba_g, aes(x,y, xend = xend,
                  yend = yend)) +
    geom_edges(col = "tomato") +
    geom_nodes(aes(size = Deg),
              alpha = 0.4 , #size = betw,
        col = "black") +
    theme_blank()
    #geom_nodetext(aes(label = vertex.names),
              #col = "black", size = 5) +
grid.arrange(ggrg, ggba, ncol = 2)
```

The Barabasi and Albert model gives us a very different degree distribution and network graph (Fig. 5.10). The Barabasi and Albert model degree distribution is positively skewed.

> ☞ **Your Turn**   Try running the simulation for the Barabasi and Albert model with n = 500. What do you observe?

5.6   Example: Electrical Automotive Goods Production Network

Amighini and Gorgoni (2014) studied the international reorganisation of auto production. Auto production uses very geographically dispersed sources for components. Data related to one of the components of auto production in 2016, electrical and electric parts, ELEnet16, is available in the ITNR package (Smith 2018).

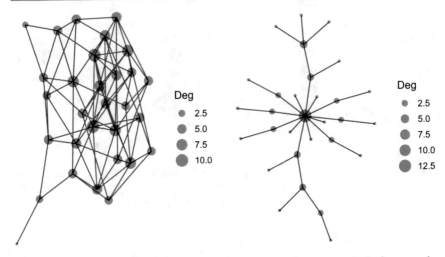

Fig. 5.10 Network diagram for Erdos Renyi (left) and Barabasi Albert (right). In the former, nodes are connected by edges randomly, and with a given probability; in the latter as the network grows, a new node is more likely to attach itself to a node that is well connected

```
library(tidygraph)
library(ITNr)
data("ELEnet16")
class(ELEnet16)
## [1] "igraph"
```

We use `tidygraph`.

```
ELE <- as_tbl_graph(ELEnet16)
class(ELE)
## [1] "tbl_graph" "igraph"
summary(ELE)
## IGRAPH a9d0bc0 DNW- 99 725 --
## + attr: name (v/c), id (v/c), regionNAME
## | (v/c), region (v/n), income (v/n), GDP
## | (v/n), GDPPC (v/n), logGDP (v/n), logGDPPC
## | (v/n), GDPgrowth (v/c), FDI (v/c), VAL
## | (e/c), Share (e/n), weight (e/n)
```

We will plot the network, but first have to do some data wrangling.

```
library(ggnetwork)
ELE_n <- asNetwork(ELE)
ELE_g <- ggnetwork(ELE_n)
str(ELE_g$regionNAME)
##  Factor w/ 7 levels "East Asia & Pacific (all income levels)",..:
                                 3 1 3 2 1 1 1 3 2 3 ...

ELE_g <- ELE_g %>%
  mutate(region_name = factor(regionNAME))

library(tidyverse)
```

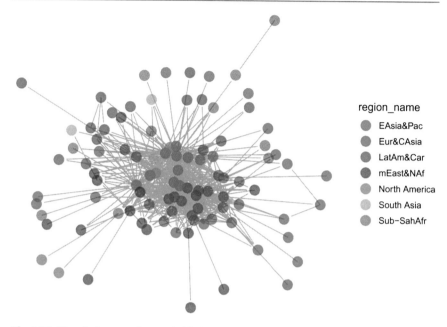

Fig. 5.11 Electrical automotive goods 2016 network

```
ELE_g <- ELE_g %>%
  mutate(region_name =
         fct_recode(region_name,
    "Sub-SahAfr" =
    "Sub-Saharan Africa (all income levels)",
    "mEast&NAf" =
    "Middle East & North Africa (all income levels)",
    "LatAm&Car" =
    "Latin America & Caribbean (all income levels)",
    "Eur&CAsia" =
    "Europe & Central Asia (all income levels)",
    "EAsia&Pac" =
    "East Asia & Pacific (all income levels)" ))


ggplot(ELE_g, aes(x,y, xend = xend,
                  yend = yend,
              col = region_name)) +
  geom_edges(color = "grey70") +
  geom_nodes(alpha = 0.6, size = 5) +
  theme_blank() +
  theme(legend.position = "right") +
  scale_colour_brewer(palette = "Dark2")
```

The auto production network diagram (Fig. 5.11) supports the observation by Amighini and Gorgoni (2014, p. 923) that auto production is "one of the most geographically fragmented activities in the manufacturing sector, with production processes split into different phases carried out in different countries." Arguing for the relevance of a network approach, they write, "Traditional approaches (such as gravitational models) consider only the relationship between countries i and j, assuming that this is independent from any other relationship i and j establish with other countries."

We now calculate a weighted outdegree measure. The edge weights are the proportion of global trade.

```
ELE2 <- ELE %>%
  activate(nodes) %>%
  mutate(outdeg = centrality_degree(weights =
         weight, mode = "out")) %>%
  as_tibble()
```

We list the 10 top countries on weighted outdegree measure (Table 5.1).

```
ELE3 <- ELE2 %>%
  select("name","regionNAME","outdeg")  %>%
  arrange(-outdeg)
library(xtable)
xtable(ELE3[1:10,], caption = "Top 10 in weighted outdegree")
```

The data for 2016 show that China, Korea and Japan were the top three countries in the valued outdegree measure. Amighini and Gorgoni (2014) estimated that the top three countries in 1998 were Germany, USA and France; in 2008 the top three countries were China, Germany and Japan. We check our calculations with the built in functions in ITNr.

```
ITN3 <- ITNcentrality(ELEnet16) %>%
  as_tibble() %>%
  arrange(-Weighted.Out.Degree)

ITN3[1:10,]
```

Table 5.1 Top 10 in weighted outdegree

	Name	regionNAME	Outdeg
1	CHN	East Asia & Pacific (all income levels)	22.92
2	KOR	East Asia & Pacific (all income levels)	10.67
3	JPN	East Asia & Pacific (all income levels)	8.74
4	USA	North America	8.45
5	MEX	Latin America & Caribbean (all income levels)	6.12
6	DEU	Europe & Central Asia (all income levels)	5.48
7	THA	East Asia & Pacific (all income levels)	2.74
8	CZE	Europe & Central Asia (all income levels)	2.35
9	MYS	East Asia & Pacific (all income levels)	2.29
10	VNM	East Asia & Pacific (all income levels)	2.21

Fig. 5.12 Weighted
outdegree distribution of
electrical automotive goods
2016 network

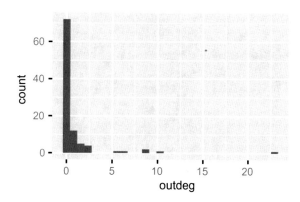

```
## # A tibble: 10 x 12
##      NAMES Weighted.Out.De~ Binary.Out.Degr~
##      <chr>            <dbl>            <dbl>
##  1 CHN                 22.9               76
##  2 KOR                 10.7               48
##  3 JPN                 8.74               32
##  4 USA                 8.45               45
##  5 MEX                 6.12               14
##  6 DEU                 5.48               46
##  7 THA                 2.74               31
##  8 CZE                 2.35               19
##  9 MYS                 2.30               14
## 10 VNM                 2.21               29
## # ... with 9 more variables:
## #   Weighted.In.Degree <dbl>,
## #   Binary.In.Degree <dbl>,
## #   Weighted.Degree.All <dbl>,
## #   Binary.Degree.All <dbl>, Betweenness <dbl>,
## #   Closeness <dbl>, Eigenvector <dbl>,
## #   Hub <dbl>, Authority <dbl>
```

We plot the weighted outdegree distribution (Fig. 5.12); it is positively skewed.

```
library(ggplot2)
ggplot(ELE2, aes(x = outdeg)) +
  geom_histogram(bins = 30)
```

We will now plot the weighted outdegree distribution by region.

```
ELE2 <- ELE2 %>%
  mutate(region_name = factor(regionNAME))

library(tidyverse)
ELE2 <- ELE2 %>%
  mutate(region_name =
         fct_recode(region_name,
    "Sub-SahAf" =
    "Sub-Saharan Africa (all income levels)",
```

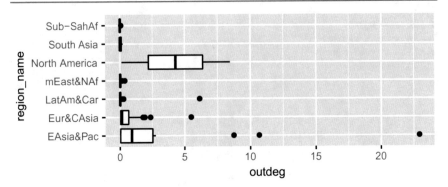

Fig. 5.13 Weighted outdegree distribution by region of electrical automotive goods 2016 network

```
    "mEast&NAf" =
    "Middle East & North Africa (all income levels)",
    "LatAm&Car" =
    "Latin America & Caribbean (all income levels)",
    "Eur&CAsia" =
    "Europe & Central Asia (all income levels)",
    "EAsia&Pac" =
    "East Asia & Pacific (all income levels)" ))

library(ggplot2)

ggplot(ELE2, aes(x = region_name,
                 y = outdeg)) +
  geom_boxplot() +
  coord_flip()
```

The region wise degree distribution shows the importance of East Asia and the Pacific, North America, and Europe and Central Asia (Fig. 5.12).

5.7   Resources

For Better Understanding

Scott Page's (undated) course on Model Thinking has very clear short lectures on networks. Barabasi (2016) has a wonderful video: Networks are everywhere. Jackson's (2014) paper links networks to Economic Behaviors. Douglas Luke (2015) has provided a detailed and clear guide to using R for network analysis

For Going Further

Jackson's (2008) book.

Your Turn Answers

☞ **Your Turn** (Sect. 5.2)   Create and plot a small network, where Tim talks to Jim, Kim talks to Jim and Jane, and Jane talks to Jim. (You can start with:

```
# install and load required packages; see chapter
# Hint
from <-  c("Kim", "Tim", "Kim", "Jim")
to   <-  c("Jim", "Jim", "Jane", "Jane")


edge2 <- tibble(from,to)


Talk2 <- tbl_graph(edges = edge2,
                     directed = FALSE)

Talk_n2 <- asNetwork(Talk2)

Talk_g2 <- ggnetwork(Talk_n2)

ggnetplot <- function(Net = Bali) {
  Net <- ggnetwork(Net, layout =
                    "kamadakawai")
  ggplot(Net, aes(x,y, xend = xend,
                   yend = yend)) +
    geom_edges(col = "tomato") +
    # node text repelled from node
    geom_nodetext_repel(aes(label = vertex.names),
            col = "black", size = 3) +
    theme_blank()
}

ggnetplot(Talk_n2)


# we are using the function created in the chapter
```

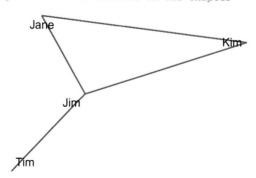

References

Amighini, A., and S. Gorgoni. 2014. The international reorganisation of auto production. *The World Economy* 37 (7): 923–952.

Barabasi, A. 2016. *Networks are everywhere.* https://www.youtube.com/watch?v=c867FlzxZ9Y. Accessed 9 Oct 2019.

Barabasi, A.-L., and R. Albert. 1999. Emergence of scaling in random networks. *Science* 286: 509–512.

Barabasi, A.-L., and E. Bonabeau. 2003. Scale-Free networks. *Scientific American* 288: 60–69.

Bojanowski, M. 2015. intergraph: Coercion routines for network data objects. *R Package Version 2.0-2.* http://mbojan.github.io/intergraph.

Briatte, F. 2016. ggnetwork: Geometries to plot networks with 'ggplot2'. *R Package Version* (5): 1. https://CRAN.R-project.org/package=ggnetwork.

Csardi, G., and T. Nepusz. 2006. The igraph software package for complex network research. *InterJournal Complex Systems* 1695: 2006. http://igraph.org.

Jackson, M.O. 2008. *Social and Economic Networks.* Princeton: Princeton University Press.

Jackson, M.O. 2014. Networks in the understanding of economic behaviors. *Journal of Economic Perspectives* 28 (4): 3–22.

Koschade, S. 2006. A social network analysis of jemaah islamiyah: the applications to counterterrorism and intelligence. *Studies in Conflict and Terrorism* 29: 559–575.

Luke, D. 2015. *A user's guide to network analysis in R.* Berlin: Springer.

Luke, D. 2018. UserNetR: data sets for a user's guide to network analysis in R. *R Package Version* 2: 26.

Padgett, J.F., and C.K. Ansell. 1993. *AJS* 98 (6): 1259–1319.

Page, S. undated. Model thinking. Coursera online course. https://www.coursera.org/. Accessed 9 Oct 2019.

Pedersen, T.L. 2019. tidygraph: A tidy API for graph manipulation. *R Package Version* 1 (1): 2. https://CRAN.R-project.org/package=tidygraph.

Schoch, D. 2017. netrankr: An R package to analyze partial rankings in networks.

Slowikowski, K. 2019. ggrepel: automatically position non-overlapping text labels with 'ggplot2'. *R Package Version* (8): 1. https://CRAN.R-project.org/package=ggrepel.

Smith, M. 2018. ITNr: Analysis of the international trade network. *R Package Version* (3): https://CRAN.R-project.org/package=ITNr.

Wickham, H. 2017. tidyverse: easily install and load the 'Tidyverse'. *R Package Version* 1 (2): 1. https://CRAN.R-project.org/package=tidyverse.

Part III
Mathematical Preliminaries for Data Analysis

Functions

6

6.1 Introduction

Functions are of great importance in economics—for example, supply and demand, and the Cobb–Douglas production function.

6.2 Making Your Own Functions in R

When we use R we use functions to get things done. We now make a few simple functions in R.

```
add2 <- function(x) {
  x + 2
}
```

We need to provide a value of x to the function add2; it will add 2 and give us an answer:

```
add2(x = 2)
## [1] 4
```

We can also specify a default value in the function:

```
mult2 <- function(M = 3) {
  2 * M
}
```

See the difference in the way add2 and mult2 work if we don't provide inputs:

© Springer Nature Singapore Pte Ltd. 2020
V. Dayal, *Quantitative Economics with R*,
https://doi.org/10.1007/978-981-15-2035-8_6

Fig. 6.1 Plot of
$y = 2 - (x + 2)^2$ with curve

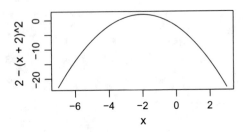

```
add2()
```

```
## Error in add2(): argument "x" is missing, with no
default
```

```
mult2()
## [1] 6
```

While making mult2 we specified a default value for M; if we don't provide an input R will use the default value.

☞ **Your Turn**   Make a function in R called Scomp for $f(S) = 2S^2 + 3S + 1$, and find f(S) for S = 333.

6.3   Plotting Functions with the Curve Function

A quick way to plot a function is to use the curve function.
If our function is $y = 2 - (x + 2)^2$, we use the following code (Fig. 6.1):

```
curve(expr = 2 - (x + 2)^2, from = -7,
      to = 3)
```

Note that we need the function to be expressed in terms of x for curve to work.
If our function is
$y = 1/(x - 2) + 3$, we plot it with (Fig. 6.2):

```
curve(expr = 1/(x-2) + 3, from = -5, to = 5)
```

☞ **Your Turn**   Use curve to plot $y = (x - 3)(x + 7)$ from $x = -15$ to $x = 15$.

Fig. 6.2 Plot of
$y = 1/(x-2) + 3$ with
curve

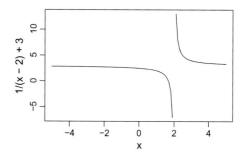

6.4   Statistical Loss Functions

We now plot two key loss functions that are used in statistics, corresponding to the mean and the median. The mean of values of a variable J can be viewed as the result of minimizing the squared loss deviations (Paolino 2017):

$$f(x) = \sum_{i=1}^{n} (J_i - x)^2$$

And the median of the values of a variable J can be viewed as the result of minimizing the absolute loss deviations (Paolino 2017):

$$f(J) = \sum_{i=1}^{n} |J_i - x|.$$

Let us consider two sets of numbers, J, and J2; J2 has one high extreme value:

```
J <- c(1,2,3,3,4)
J2 <- c(1,2,3,3,9)

mean(J)
## [1] 2.6
median(J)
## [1] 3
mean(J2)
## [1] 3.6
median(J2)
## [1] 3
```

We plot the loss function corresponding to the mean for J and J2 (Fig. 6.3); the curve is lowest at the mean:

```
curve((J[1] - x)^2 + (J[2] - x)^2 +
        (J[3] - x)^2 + (J[4] - x)^2 +
        (J[5] - x)^2, 1, 7,
      ylab = "Loss")
curve((J2[1] - x)^2 + (J2[2] - x)^2 +
        (J2[3] - x)^2 + (J2[4] - x)^2 +
        (J2[5] - x)^2, 1, 7,
      add = TRUE, lty = 2, ylab = "Loss")
```

We plot the loss function corresponding to the median for J and J2 (Fig. 6.4); the curve is lowest at the median:

Fig. 6.3 Loss function for mean, J2 is dashed line, J is solid line

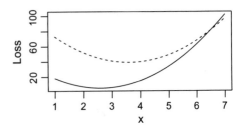

Fig. 6.4 Loss function for median, J2 is dashed line, J is solid line

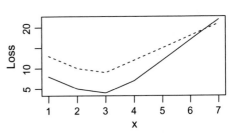

```
curve(abs(J[1] - x) + abs(J[2] - x) +
      abs(J[3] - x) + abs(J[4] - x) +
      abs(J[5] - x), 1, 7,
   ylab = "Loss")
curve(abs(J2[1] - x) + abs(J2[2] - x) +
      abs(J2[3] - x) + abs(J2[4] - x) +
      abs(J2[5] - x), 1, 7,
      add = TRUE, lty = 2, ylab = "Loss")
```

6.5   Supply and Demand

We will now plot hypothetical demand and supply curves (Fig. 6.5):

- An initial inverse demand curve: $p_D = (125 - 6q)/8$
- A shifted inverse demand curve: $p_D = (150 - 6q)/8$
- A supply curve: $p_S = (12 + 2q)/5$

```
curve((125 - 6*x)/8, 0,30,
      ylim=c(0,20))
curve((150 - 6*x)/8, 0,30,
      lty = 2, add = TRUE)
curve((12 + (2 * x ))/5,
      add = TRUE)
```

Supply and demand can be useful in interpreting global oil prices. We will now download oil price data from: https://www.bp.com/en/global/corporate/energy-economics/statistical-review-of-world-energy.html, and plot it, using the tidyverse package (Wickham 2017).

Fig. 6.5 Supply and demand

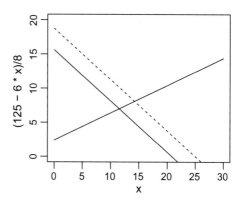

```
library(tidyverse)
Oil <- read_csv("Oil.csv")
```

```
## Parsed with column specification:
## cols(
##   Year = col_double(),
##   Price_nominal = col_double(),
##   Price_2018 = col_double()
## )
```

We plot the price of oil (Fig. 6.6):

```
ggplot(Oil, aes(x = Year, y = Price_2018)) + geom_line() +
  scale_x_continuous(limits = c(1861,2018),
        breaks = seq(1860, 2020, by = 20))
```

The 1979 oil shock was caused by a reduction in output related to political events in Iran. There was a rise in prices from 1999 till mid-2008 because of rising oil demand in countries like China and India.

☞ **Your Turn**   Look up the following websites and relate them to the graph of oil prices. Wikipedia article on price of oil and https://www.winton.com/longer-view/price-history-oil.

6.6   Cobb–Douglas Production Function

A Cobb–Douglas production function has the form $f(x_1, x_2) = A x_1^a x_2^b$.
 We will use the mosaic package (Pruim et al. 2017) for plotting. We take specific values of the parameters as in the code below. We plot the isoquants (Fig. 6.7):

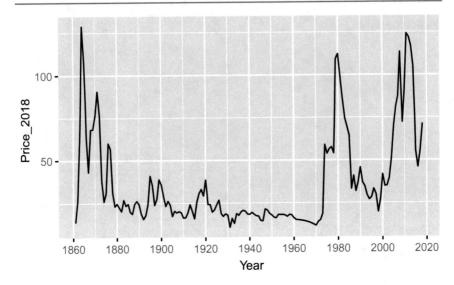

Fig. 6.6 Price of oil

Fig. 6.7 Cobb–Douglas
production function

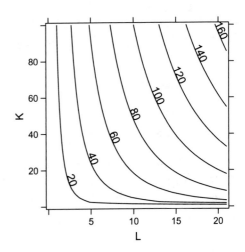

```
library(mosaic)
plotFun(A * (L ^ 0.7) * (K ^ 0.3) ~ L & K,
        A = 5, filled = FALSE,
        xlim = range(0, 21),
        ylim = range(0, 100))
```

The following code gives us a three-dimensional view (Fig. 6.8):

```
library(mosaic)
plotFun(A * (L ^ 0.7) * (K ^ 0.3) ~ L & K,
        A = 5, filled = FALSE,
        xlim = range(0, 21),
```

Fig. 6.8 Cobb–Douglas
production function

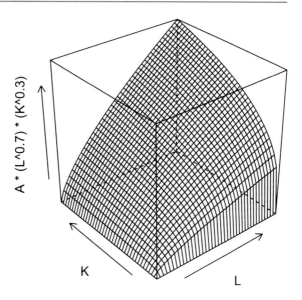

```
         ylim = range(0, 100),
         surface = TRUE)
```

We will now work with data in the `micEcon` package (Henningsen 2017). The package contains data on 140 French apple producers for the year 1986.

```
data( "appleProdFr86", package = "micEcon" )
dat <- appleProdFr86
rm( appleProdFr86 )
```

In the data Cap denotes capital (including land); Lab labour; and Mat intermediate materials. Input quantities (q) are calculated from the costs (v) and price indexes (p):

```
dat$qCap <- dat$vCap / dat$pCap
dat$qLab <- dat$vLab / dat$pLab
dat$qMat <- dat$vMat / dat$pMat
```

We make scatter plots of log of output versus log of inputs (Figs. 6.9, 6.10 and 6.11):

```
ggplot(dat, aes(y = log(qOut), x = log(qCap))) + geom_point() +
  geom_smooth(method = "lm")


ggplot(dat, aes(y = log(qOut), x = log(qLab))) + geom_point() +
  geom_smooth(method = "lm")


ggplot(dat, aes(y = log(qOut), x = log(qMat))) + geom_point() +
  geom_smooth(method = "lm")
```

Fig. 6.9 Scatter plot of log of output versus log of capital

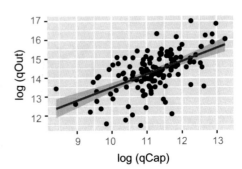

Fig. 6.10 Scatter plot of log of output versus log of labour

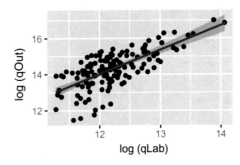

Fig. 6.11 Scatter plot of log of output versus log of materials

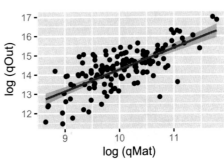

We estimate a Cobb–Douglas production function and tabulate with the `texreg` package (Leifeld 2013) (results in Table 6.1).

```
prodCD <- lm( I(log( qOut )) ~ log( qCap ) + log( qLab ) +
            log( qMat ), data = dat )
library(texreg) texreg(list(prodCD), caption = "Dependent variable
is log(qOut)",
        caption.above = TRUE)
```

The coefficients give us the output elasticities with respect to the inputs. If an apple producer increases capital by one percent, the output will increase by 0.16%. Increases in labour and materials inputs by one percent will lead to estimated increases of 0.68% and 0.63%, respectively.

Table 6.1 Dependent variable is log(qOut)

	Model 1
(Intercept)	−2.06
	(1.31)
log(qCap)	0.16
	(0.09)
log(qLab)	0.68***
	(0.15)
log(qMat)	0.63***
	(0.13)
R^2	0.59
Adj. R^2	0.59
Num. obs.	140
RMSE	0.66

$^*p < 0.05; ^{**}p < 0.01; ^{***}p < 0.001$

6.7   Resources

For Better Understanding

Varian (2014) is clear and intuitive. Klein's (1962) classic text has much to offer the applied economist.

For Going Further

Henningsen's (2018) lecture notes on economic production analysis are detailed and use R.

Your Turn Answers

☞ **Your Turn (Sect. 6.2)**   Make a function in R called Scomp for $f(S) = 2S^2 + 3S + 1$, and find f(S) for S = 333.

```
# Make function
Scomp <- function(S = 2) {
  (2 * (S^2)) + (3 * S) + 1
}
# check
Scomp(S = 2)
## [1] 15
# get answer you want
Scomp(S = 333)
## [1] 222778
```

☞ **Your Turn (Sect. 6.3)**   Use curve to plot $y = (x - 3)(x + 7)$ from $x = -15$ to $x = 15$.

```
curve(expr = (x - 3) * (x + 7), from = - 10, to = 10)
```

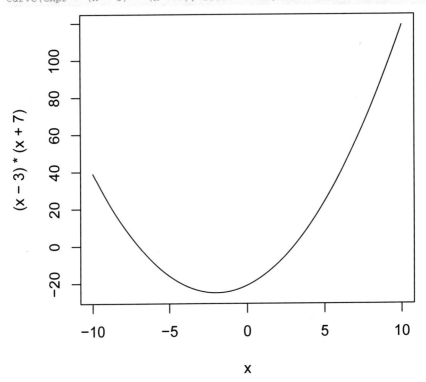

References

Henningsen, A. 2017. micEcon: microeconomic analysis and modelling. R package version 0.6–14. https://CRAN.R-project.org/package=micEcon

Henningsen, A. 2018. *Introduction to econometric production analysis with R*. Collection of lecture notes. 2nd draft version. Department of Food and Resource Economics, University of Copen-hagen. http://leanpub.com/ProdEconR/

Klein, L.R. 1962. *An introduction to econometrics*. Englewood Cliffs: Prentice-Hall.

Leifeld, P. 2013. texreg: conversion of statistical model output in R to LaTeX and HTML tables. *Journal of Statistical Software* 55 (8): 1–24. http://www.jstatsoft.org/v55/i08/

Paolino, J. 2017. Teaching univariate measures of location–using loss functions. *Teaching Statistics Trust* 40 (1): 16–23.

Pruim, R., D.T. Kaplan, and N.J. Horton. 2017. The mosaic package: helping students to 'think with data' using R. *The R Journal* 9 (1): 77–102.

Varian, H. 2014. *Intermediate economics with calculus*. New York: W W Norton and Company.

Wickham, H. 2017. tidyverse: easily install and load the 'tidyverse'. R package version 1.2.1. https://CRAN.R-project.org/package=tidyverse

Difference Equations

7

7.1　Introduction

We often model how quantities change over time using difference equations. Today's level of a variable will be related to yesterday's level of the variable and also other variables.

7.2　Simple Toy Example

```
library(tidyverse)
```

We consider a simple difference equation for a variable X:

$X_t = X_{t-1} + 2$ where t denotes time.

Let the initial value of X be:

$X_0 = 10$

Then $X_1 = 10 + 2 = 12$, and $X_2 = 12 + 2 = 14$. We can calculate the value of X for different time periods using a loop (Fig. 7.1):

```
X <- numeric(10)
X[1] = 10
for(i in 2:10){
  X[i] <- X[i -1] + 2
}
X
## [1] 10 12 14 16 18 20 22 24 26 28
Time <- 1:10
X_time <- tibble(X, Time)
ggplot(X_time, aes(x = Time, y = X)) +
  geom_line()
```

© Springer Nature Singapore Pte Ltd. 2020
V. Dayal, *Quantitative Economics with R*,
https://doi.org/10.1007/978-981-15-2035-8_7

Fig. 7.1 X versus time

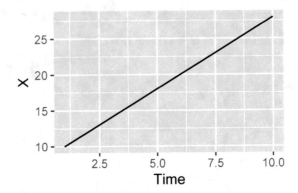

> ☞ **Your Turn** Simulate and plot the following difference equation:
> $X_t = 0.5 * X_{t-1} + 5$, with $X_0 = 12$

7.3   Example: Global Carbon Stocks

Common and Stagl (2005, p. 502) use a simple difference equation to illustrate how global carbon stocks may change over time; it is an approximation, but a useful one.

We use S to denote the stock of carbon, E to denote emissions and t denote time. Then

$S_t = S_{t-1} + E_t - dS_{t-1}$

where dS_{t-1} is how much of the stock is absorbed.

The atmospheric stock of carbon is about 750 Gt. From data, the parameter d can be approximated by 0.005. If current emissions continue at the same level for the next 500 years, we have the following scenario:

```
S <- numeric(500)
E <- numeric(500)
S[1] <- 750
E[1] <- 6.3

for(i in 2:500) {
  E[i] <- E[i-1]
  S[i] <- S[i-1] + E[i-1] - (0.005*S[i-1])
}
Time <- 1:500
C_scenario <- tibble(S,Time,E)
```

We plot S versus time (Fig. 7.2); S increases and then stabilizes.

Fig. 7.2 S versus time

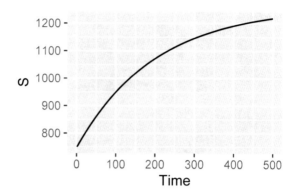

```
ggplot(C_scenario, aes(x = Time,
                    y = S)) +
  geom_line()
```

We now write a function where we can incorporate different growth rates of emissions over the next 50 years. If emissions grow at 1.4% a year over the next 50 years and then stay at that level for the next 450 years, we have the following scenario:

```
CO2 <- function(growth = 1.014) {
S <- numeric(500)
E <- numeric(500)
S[1] <- 750
E[1] <- 6.3
for(i in 2:50) {
  E[i] <- E[i-1] * growth
  S[i] <- S[i-1] + E[i-1] - (0.005*S[i-1])
}
for(i in 51:500) {
  E[i] <- E[i-1]
  S[i] <- S[i-1] + E[i-1] - (0.005*S[i-1])
}
Time <- 1:500
C_scenario <- tibble(S,Time,E)

ggplot(C_scenario, aes(x = Time,
  y = S)) +
  geom_line()
}
```

Fig. 7.3 S versus time

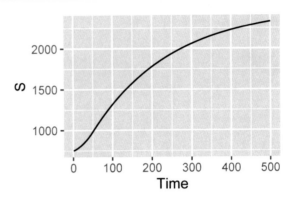

Now that we have written the function, we can simulate and generate a plot easily (Fig. 7.3):

```
CO2(1.014)
```

7.4   Fish

We use difference equations to see the stock, growth and harvest of fish over time.

7.4.1   Numerical Simulation

We use x to denote fish stock, G to denote fish growth and H to denote harvest.
$$x_t = x_{t-1} + G_{t-1} - H_{t-1}$$
G follows a logistic function:
$$G_{t-1} = rx_{t-1}[1 - (x_{t-1}/K)]$$
H is a fraction h of the fish stock:
$$H_{t-1} = hx_{t-1}$$
We make a function called logistic to simulate and plot G, H and x.

```
logistic <- function(r = 0.05, xinit = 40,
                     h = 0.02) {
  K <- 100
  x <- numeric(101)
  gth <- numeric(101)
  H <- numeric(101)
  x[1] <- xinit
  gth[1] <- 0
  for (i in 2:101) {
    xbyk <- x[i - 1]/K
    gth[i - 1] <- r*x[i - 1]*(1 - xbyk)
```

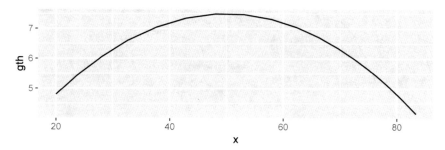

Fig. 7.4 Fish variables over time

```
  H[i - 1] <- h*x[i - 1]
  x[i] <- x[i - 1] + gth[i - 1] - H[i - 1]
}
library(tidyverse)
Time <- 1:101
xtab <- tibble(gth,H,x,Time)
xtab <- xtab[1:100,]

gg1 <- ggplot(xtab, aes(y = gth,
                        x = x)) +
  geom_line()
gg2 <- ggplot(xtab, aes(y = H,
                        x = Time)) +
  geom_line()
gg3 <- ggplot(xtab, aes(y = x,
                        x = Time)) +
  geom_line()
mylist <- list(gg1,gg2,gg3)
return(mylist)
}
```

We can run the function and get a plot (Figs. 7.4, 7.5 and 7.6); all we have to do is provide inputs for the intrinsic rate of growth r, the initial value of x, xinit and the fraction of stock harvested, h:

```
logistic(r = 0.3, xinit = 20, h = 0.05)
## [[1]]

##
## [[2]]

##
## [[3]]
```

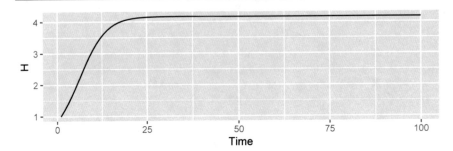

Fig. 7.5 Fish variables over time

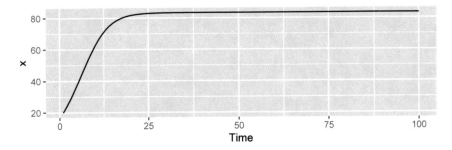

Fig. 7.6 Fish variables over time

☞ **Your Turn**   Copy and paste the code for the logistic function above (or write your own function), and simulate and plot for values: r = 2.6, xinit = 20 and h = 0.05.

7.4.2   Example: North Sea Herring

Bjorndal and Conrad (1987) modelled the open access exploitation of North Sea herring during the period 1963–77. Fish populations are often modelled with a logistic equation.

Denoting the fish stock with S, the intrinsic rate of growth by r and its carrying capacity by L, we have: $S_t = S_{t-1} + r S_{t-1}(1 - (S_{t-1}/L))$

If $S_{t-1} = 0$ or L, the growth is zero. Growth in fish depends on the stock of fish. We first see how S changes over time if the fish are left alone, i.e. there is no harvest.

We need to give R the formula, the values for r and L, and the value for initial S.

```
S = numeric(15)
S[1] = 2325000
r = 0.8; L = 3200000
```

Fig. 7.7 S versus time

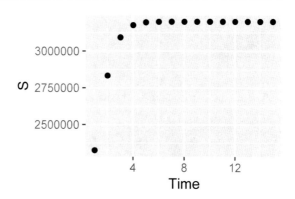

We use a loop as before, within the loop we use the formula for logistic growth to calculate the stock S in each period.

```
for(t in 2:15){
    S[t] <- S[t-1]+(S[t-1]*r)*(1-S[t-1]/L)
}
Time <- 1:15
```

We now plot S against Time (Fig. 7.7):

```
SandTime <- data.frame(S,Time)
ggplot(SandTime, aes(x = Time, y = S)) +
    geom_point()
```

The fish stock increases until it is equal to the carrying capacity (Fig. 7.7).

We now examine the fish stock with fishing. Fishing harvest depends on the fish stock S2 and fishing capital, K. Bjorndal and Conrad (1987) use a Cobb–Douglas production function. We will not go into the details of their derivation, but they arrive at the following dynamic system that they use for a simulation:

$$K_{t+1} = K_t + n(aK_t^{b-1}S_t^g - c_t/p_t)$$
$$S_{t+1} = S_t + rS_t(1 - S_t/L) - aK_t^b S_t^g$$

The equation on the top represents the adjustment of capital to profit—higher profits lead to an expansion of capital. We had used the lower equation earlier; it represents the biological growth and harvest of fish. We now have more parameters that we have to provide numerical values for:

```
S2 = numeric(15)
K = numeric(15)
S2[1] = 2325000
K[1] = 120
r = 0.8; L = 3200000;
a = 0.06157; b = 1.356; g = 0.562; n = 0.1
c <- c(190380,195840,198760,201060,204880,206880,
```

```
        215220,277820,382880,455340,565780,686240,556580,
        721640,857000)
p <- c(232,203,205,214,141,128,185,262,244,214,
        384,498,735,853,1415)
```

We put the two equations into a loop:

```
for(t in 2:15){
  S2[t] <- S2[t-1]+(S2[t-1]*r)*(1-S2[t-1]/L) -
    a*K[t-1]^b*S2[t-1]^g
  K[t] <- K[t-1]+( n*(a*( K[t-1]^(b-1) )* (S2[t-1]^g) -c[t-1]
                     /p[t-1]))
}
```

We plot K against S2

```
KandS2 <- data.frame(K,S2)
ggplot(KandS2, aes(x = S2, y = K)) +
  geom_path(arrow = arrow())
```

and S against Time

```
S2andTime <- data.frame(S2, Time)

ggplot(S2andTime, aes(x = Time, y = S2)) +
  geom_line()
```

We see that with open access fishing the dynamics of the fishery is dramatically changed from the case of no fishing. In Fig. 7.8 we see that initially fishing leads to an expansion of capital which over time leads to a reduction in the fishing stock. When capital is around 600, it starts falling but the reduction in capital is a case of too little, too late. We can contrast Fig. 7.9 with Fig. 7.7; in Fig. 7.9 the fish die out.

Fig. 7.8 K versus S2

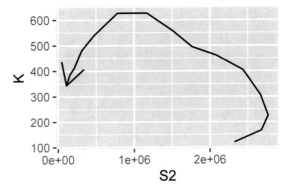

Fig. 7.9 S2 versus time

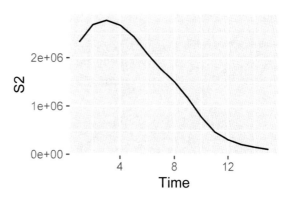

In their conclusions Bjorndal and Conrad (1987, pp. 83–84) write:

> In the empirical analysis of open access systems it is important to note that non-linear differ-
> ence equations, with or without longer lags, are capable of more complex dynamic behaviour
> than their continuous-time (differential equation) analogues. The lag in adjustment by both
> the exploited species and the harvesters themselves is often a more accurate depiction of
> dynamics, and the differential equation systems are best viewed as theoretical approxima-
> tions. ... If adjustment in an open access system is discrete, there is a greater likelihood of
> overshoot, severe depletion, and possibly extinction. ... the North Sea herring fishery ... was
> saved from more severe overfishing and possibly extinction by the closure of the fishery at
> the end of the 1977 season.

7.5 Example: Conrad's Model of a Stock Pollutant

Economists have long advocated using economic instruments to tackle pollution. A
wide variety of applications exist, and theoretical models are the starting points for
understanding such applications, and the issues that arise. Conrad (2010) presents a
model of a stock pollutant and emission taxes.

Wastes may accumulate over time, and this is what we call a stock pollutant. In
this case, the economic issue includes dynamic considerations.

Using numerical values and specific functional forms can make models more
intuitive.

7.5.1 Commodity Residual Transformation Function

In this model we have an industry that produces commodities, and wastes are produced along with these commodities.

We have an implicit *commodity–residual transformation function* that each firm
in the industry faces:

$$\phi(Q_t, S_t) = 0 \qquad (7.1)$$

Fig. 7.10 Commodity
(Q)–residual (S)
transformation function

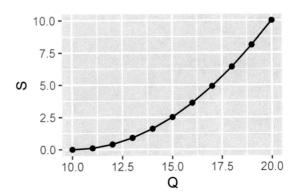

where Q is the commodity, S is the waste, and t is the time.

We can use a specific form for the commodity–residual transformation function.

$$(Q_t - m^2) - nS_t = 0 \tag{7.2}$$

We will assume certain values for m and n, and then plot the commodity–residual transformation function (Fig. 7.10).

```
library(tidyverse)
m <- 10
n <- 10
Q <- 10:20
S <- ((Q - m)^2)/m
qplot(Q,S) + geom_line()
```

7.5.2 Stock Pollutant

The accumulation of the stock pollutant Z is given by the difference equation:

$$Z_{t+1} - Z_t = -\gamma Z_t + N S_t \tag{7.3}$$

Once again we can assume some numerical values and plot (Fig. 7.11).

```
TimeSpan <- 20
Z <- numeric(TimeSpan)
Z[1] <- 1400
gamma <- 0.2
S <- 3
N <- 100

N * S / gamma
```

Fig. 7.11 Waste accumulation, Z is stock pollutant

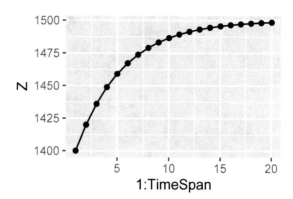

```
## [1] 1500
for (i in 2:TimeSpan) {
Z[i] <- Z[i-1] - (gamma * Z[i-1]) +
   (N * S)
}

qplot(1:TimeSpan,Z) + geom_line()
```

7.5.3   Firm's Choice of Commodity Q Given a Tax on Waste S

The revenue of the price-taking firm in each period is pQ_t while the cost of emitting waste is $\tau_t S_t$, and the firm is assumed to maximize its revenue in each period given the commodity residual transformation function. So the firm works through the maximization every period and solves this static problem.

Assuming the specific form for the commodity residual transformation function, we get the following solution to the firm's problem:

$$Q_t = (np/(2\tau_t)) + m \tag{7.4}$$

```
p <- 200
m <- 10
Q_static <- numeric(20)
tau <- seq(from = 100, to = 400,
          length.out = 20)
Q_static <- (n*p/(2*tau)) + m
qplot(tau, Q_static) + geom_line()
```

This is plotted in Fig. 7.12.

Fig. 7.12 Firm's optimal Q
for different tax rates tau

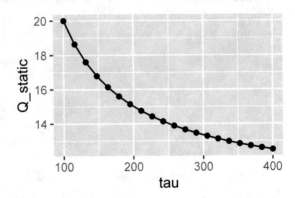

7.5.4   What Is the Optimal Tax?

The regulator can choose tau to influence Q; so the regulator sees what the optimal
Q is. The regulator solves the following problem:

Maximize $\sum_{t=0}^{infty} \rho^t (pNQ_t - cZ_t^2)$, subject to constraints.
The optimal steady state Q is solved as:

$$Q^* = \sqrt[3]{\frac{n^2 p(\delta + \gamma)\gamma}{4cN}} + m \tag{7.5}$$

As before, we can plot the optimal Q and tau as functions of parameters, for
example, delta (δ), which is the discount rate.

```
n <- 10
p <- 200
delta <- c(0.025, 0.05,0.075,
           0.1, 0.125)
gamma <- 0.2
c <- 0.02
N <- 100
m <- 10

Num <- (n^2) * p * (delta + gamma) *
   gamma
Denom <- 4 * c * N
Qfrac <- (Num  / Denom )^(1/3)
Qstar <- Qfrac + m
Qstar
## [1] 14.82745 15.00000 15.16140 15.31329 15.45696
qplot(delta,Qstar) + geom_line()
```

Fig. 7.13 Optimal Q, Qstar for different deltas

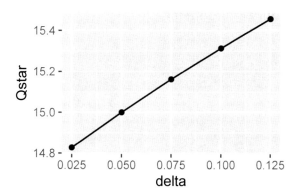

Fig. 7.14 Optimal tau, taustar, for different deltas

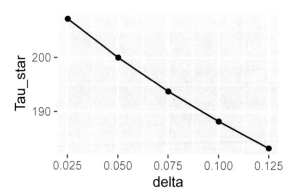

```
Tau_star <- n*p/(2*(Qstar - m))
qplot(delta, Tau_star) + geom_line()
```

See Figs. 7.13 and 7.14. Qstar increases with delta and Tau_star decreases with delta.

7.6   Resources

For Better Understanding

Common and Stagl (2005) is an accessible but clear text on Ecological Economics.

For Going Further

Conrad (2010) uses numerical examples in Excel to discuss the intuition behind complicated resource economics models.

Your Turn Answers

☞ **Your Turn (Sect. 7.2)**   Simulate and plot the following difference equation:
$$X_t = 0.5 * X_{t-1} + 5, \text{ with } X_0 = 12$$

```
X <- numeric(10)
X[1] = 12
for(i in 2:10){
  X[i] <- 0.5 * X[i -1] + 5
}
X
##  [1] 12.00000 11.00000 10.50000 10.25000 10.12500
##  [6] 10.06250 10.03125 10.01562 10.00781 10.00391
Time <- 1:10
X_time <- tibble(X, Time)
ggplot(X_time, aes(x = Time, y = X)) +
  geom_line()
```

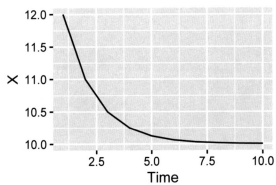

☞ **Your Turn (Sect. 7.4)**   Copy and paste the code for the logistic function above (or write your own function), and simulate and plot for values: r = 2.6, xinit = 20 and h = 0.05 (Figs. 7.15, 7.16 and 7.17).

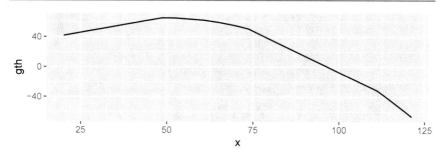

Fig. 7.15 Fish variables over time

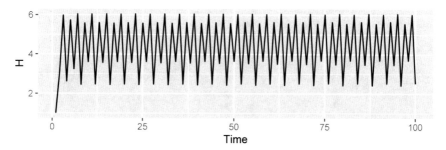

Fig. 7.16 Fish variables over time

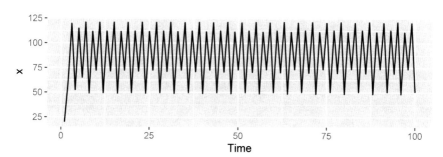

Fig. 7.17 Fish variables over time

```
logistic <- function(r = 0.05, xinit = 40,
                      h = 0.02) {
  K <- 100
  x <- numeric(101)
  gth <- numeric(101)
  H <- numeric(101)
  x[1] <-  xinit
  gth[1] <- 0
  for (i in 2:101) {
    xbyk <- x[i - 1]/K
```

```
  gth[i - 1] <- r*x[i - 1]*(1 - xbyk)
  H[i - 1] <- h*x[i - 1]
  x[i] <- x[i - 1] + gth[i - 1] - H[i - 1]
}
library(tidyverse)
Time <- 1:101
xtab <- tibble(gth,H,x,Time)
xtab <- xtab[1:100,]
#xtab2 <- tibble(x,Time) not needed
gg1 <- ggplot(xtab, aes(y = gth,
                        x = x)) +
  geom_line()
gg2 <- ggplot(xtab, aes(y = H,
                        x = Time)) +
  geom_line()
gg3 <- ggplot(xtab, aes(y = x,
                        x = Time)) +
  geom_line()
mylist <- list(gg1,gg2,gg3)
return(mylist)
}
```

$r = 2.6$, xinit $= 20$ and h $= 0.05$.

```
logistic(r = 2.6, xinit = 20, h = 0.05)
## [[1]]

##
## [[2]]

##
## [[3]]
```

References

Bjorndal and Conrad. 1987. The dynamics of an open access fishery. *Canadian Journal of Economics* 20 (1): 74–85.

Common, M., and S. Stagl. 2005. *Ecological economics*. Cambridge: Cambridge University Press.

Conrad, J. 2010. *Resource economics*, 2nd ed. Cambridge: Cambridge University Press.

Matrices

8

8.1 Introduction

Moore and Siegel (2013, p. 275) explain that we use different ways—matrices, vectors and scalars—to represent variables:

> A matrix holds more information than a vector, which holds more information than a scalar. ... The key point to appreciate is that scalar, vector, and matrix algebra define the rules for manipulating variables in different representations.

8.2 Simple Statistics with Vectors

We start with a vector x, which has the following elements:

```
x <- c(1,2,3,4,5)
```

We can find the length of x, the sum of x and the transpose of x:

```
length(x)
## [1] 5
sum(x)
## [1] 15
t(x) # transpose of x
##      [,1] [,2] [,3] [,4] [,5]
## [1,]    1    2    3    4    5
```

© Springer Nature Singapore Pte Ltd. 2020
V. Dayal, *Quantitative Economics with R*,
https://doi.org/10.1007/978-981-15-2035-8_8

Let us consider a vector of ones, of the same length as x:

```
ones <- c(rep(1,length(x)))

length(ones)
## [1] 5
sum(ones)
## [1] 5
t(ones)
##        [,1] [,2] [,3] [,4] [,5]
## [1,]    1    1    1    1    1
```

Now consider what happens when we multiply the transpose of x by ones, using the operator for vector or matrix multiplication in R (%*%):

```
t(x) %*% ones
##        [,1]
## [1,]    15
sum(x)
## [1] 15
```

What we did above via vector multiplication is summation.
The mean of x, $\bar{x}$ can be computed:

```
t(x) %*% ones / length(x)        .
##        [,1]
## [1,]     3
mean(x)
## [1] 3
```

The formula for the variance of x is $var_x = [1/(n-1)] \sum_{i=1}^{n} (x_i - \bar{x})^2$; involving more involved summation than for the mean.

We first calculate a vector of the deviation of each value of x from the mean of x, square the deviations, add them up, and then divide by n−1.

We get the deviations of values of x from their mean:

```
dev.x <- x-mean(x); dev.x
## [1] -2 -1  0  1  2
```

To square these values and add the squared values we multiply the transpose of dev.x by dev.x. And to repeat, we use not star but % * %.

```
t(dev.x) %*% dev.x
##      [,1]
## [1,]   10
```

We now calculate var_x, the variance of x:

```
Var_calc_x <- t(dev.x) %*% dev.x/(length(x) - 1)
Var_calc_x
##      [,1]
## [1,]  2.5
```

We check that the variance of x has been calculated correctly by using the R function for variance, `var`:

```
var(x)
## [1] 2.5
```

8.3 Matrix Operations

We use `matrix` to make a matrix:

```
# ncol is number of columns
A <- matrix(c(4,3,6,4),ncol=2); A
##      [,1] [,2]
## [1,]    4    6
## [2,]    3    4
B <- matrix(c(2,5,6,1),ncol=2); B
##      [,1] [,2]
## [1,]    2    6
## [2,]    5    1
```

We can add the matrices A and B.

```
M <- B + A
M
##      [,1] [,2]
## [1,]    6   12
## [2,]    8    5
```

The transpose of a matrix switches the rows and columns:

```
t(A)
##        [,1] [,2]
## [1,]    4    3
## [2,]    6    4
t(B)
##        [,1] [,2]
## [1,]    2    5
## [2,]    6    1
```

We can represent a set of equations compactly with matrices; for example, $4w + 6z = 14$ and $3w + 4z = 10$. We can represent this with matrices as $AD = C$. Here D is a column vector with elements w and z and C is also a column vector with elements 14 and 10. Since $AD = C$, $D = A^{-1}C$. We invert matrices by using 'solve' in R.

```
C = c(14,10)
D <- solve(A) %*% C
D
##        [,1]
## [1,]    2
## [2,]    1
A %*% D
##        [,1]
## [1,]    14
## [2,]    10
```

Hence, $w = 2$ and $z = 1$.

8.4   Example: Poverty Rate and Relative Income

We now work with an example with actual data reported in Gill (2006), concerning the poverty rate and relative income of citizens over 65 years of age in different European Union (EU) member countries in 1998. There are two variables for those over 65, relative income, and poverty rate (Gill 2006, p. 156): '(1) the median (EU standardized) income of individuals aged 65 and older as a percentage of the population age 0–64, and (2) the percentage with income below 60% of the median (EU standardized) income of the national population'.

```
Poverty_rate <- c(7, 8, 8, 11, 14, 16, 17, 19,
    21, 22, 24, 31, 33, 33, 34)
Relative_income <- c(93, 99, 83, 97, 96, 91, 78, 90,
   78, 76, 84, 68, 76, 74, 69)
```

We create related vectors that we call Y and One, and a matrix X.

```
Y <- Poverty_rate
One <- c(rep(1, length(Y)))
X <- cbind(One, Relative_income)
```

We calculate the mean and variance of relative income:

```
RI <- Relative_income
t(RI) %*% One / length(RI)
##          [,1]
## [1,] 83.46667
mean(RI)
## [1] 83.46667
```

```
var(RI)
## [1] 105.8381
dev.RI <- RI - mean(RI)
Var_calc_RI <- t(dev.RI) %*% dev.RI/(length(RI) - 1);Var_calc_RI
##          [,1]
## [1,] 105.8381
```

☞ **Your Turn**   Use matrix operations to calculate the mean and variance of the poverty rate.

We would like to run a regression of the poverty rate on relative income.

```
lm(Poverty_rate ~ Relative_income)
##
## Call:
## lm(formula = Poverty_rate ~ Relative_income)
##
## Coefficients:
##      (Intercept)    Relative_income
##          83.6928             -0.7647
```

```
data_eu <- data.frame(Poverty_rate = Poverty_rate,
  Relative_income = Relative_income)
library(ggplot2)
ggplot(data_eu, aes(x = Relative_income, y = Poverty_rate)) +
  geom_point() +
  geom_smooth(method = "lm", se = FALSE)
```

The poverty rate is negatively related to relative income (Fig. 8.1).

We can also use the matrix formula for least squares to estimate the regression coefficients: $Coeff = (X^T X)^{-1} X^T Y$.

```
matcoeff <- solve(t(X) %*% X) %*% t(X) %*% Y
matcoeff
##                        [,1]
## One              83.69279
## Relative_income  -0.76469
```

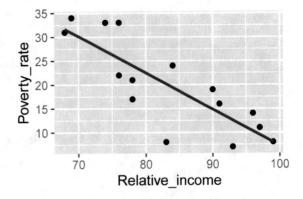

Fig. 8.1 Scatter plot of poverty rate versus relative income

8.5   Resources

For Better Understanding

The books by Gill (2006) and Moore and Siegel (2013) are wonderfully written. Namboodiri (1984) provides a detailed and patient exposition.

For Going Further

Sydsaeter and Hammond (1995) is mathematically rigorous, is written for economists and is a good text to get practice with more advanced topics.

Your Turn Answers

Use matrix operations to calculate the mean and variance of the poverty rate.

```
PR <- Poverty_rate
t(PR) %*% One / length(PR)
##             [,1]
## [1,] 19.86667
mean(PR)
## [1] 19.86667



dev.PR <- PR - mean(PR)
Var_calc_PR <- t(dev.PR) %*% dev.PR/(length(PR) - 1)
Var_calc_PR
##             [,1]
## [1,] 91.12381
var(PR)
## [1] 91.12381
```

References

Gill, J. 2006. *Essential mathematics for political and social research.* New York: Cambridge University Press.

Moore, W.H., and D.A. Siegel. 2013. *A mathematics course for political and social research.* Princeton: Princeton University Press.

Namboodiri, K. 1984. *Matrix algebra: an introduction.* London: Sage Publications Ltd.

Sydsaeter, K., and P.J. Hammond. 1995. *Mathematics for economic analysis.* Delhi: Pearson Education Inc.

Part IV
Inference from Data

Statistical Inference

9

9.1 Introduction

We will use computer simulation to illustrate key ideas in statistical inference and also explore key computer-based methods—the bootstrap and permutation methods. According to Efron and Tibshirani (1991, p. 390), 'Modern electronic computation has encouraged a host of new statistical methods that require fewer distributional assumptions than their predecessors and can be applied to more complicated statistical estimators. These methods allow the scientist to explore and describe data and draw valid statistical inferences without the usual concerns for mathematical tractability'. The numbers generated in each simulation are different.

9.2 Box Models

Freedman, Pisani and Purves (2009) use a box model to illustrate concepts of statistical inference. We can think of a random variable as numerical outcomes that result from draws of a ticket from a box.

Tossing a coin is like drawing from a box that has two tickets in it, one labelled H for heads and another labelled T for tails.

9.2.1 Sample

If the box is $[H, T]$ and I close my eyes and randomly draw a ticket, the drawn ticket could be H or T with equal probability. Suppose the ticket drawn is H. If I do not replace the ticket, then the box is now $[T]$ and if I close my eyes and randomly draw a ticket, then I will draw T with certainty. If, however, I replace H in the box, then the box is once again $[H, T]$ and if I close my eyes and randomly draw a ticket,

© Springer Nature Singapore Pte Ltd. 2020
V. Dayal, *Quantitative Economics with R*,
https://doi.org/10.1007/978-981-15-2035-8_9

it could be H or T with equal probability. So if I draw tickets twice in succession then the second draw is dependent on the first if I do not replace the ticket, and is independent of the first if I replace the ticket. Replacing the ticket is like having a box with a huge number of tickets. If the box has a huge number of H and T, and we draw twice, then the second draw is effectively independent of the first.

We create this box in R below and call it `Coin_box`.

```
library(tidyverse)
rm(list = ls()) # remove previous objects
Coin_box <- c("H", "T")
Coin_box
## [1] "H" "T"
```

If we flip a coin, we will either get a head or a tail. If we flip a coin three times, we get different sequences of heads and tails. We can simulate the flipping of a coin by using R. We use the sample function and draw tickets from Coin box.

```
# Flipping coins in the Coin_box
sample(x = Coin_box, size = 2,
       replace = TRUE)
## [1] "T" "T"
```

The arguments of the sample function are:

1. the object to be sampled from,
2. the size of the sample and
3. whether to replace or not.

If replace is TRUE, then the observations in each sample are independent; it is like sampling from an infinitely large population. The default value is FALSE. With replace = FALSE, if we draw a head ticket from Coin box the first time, the next draw will be of a tail. And vice versa. The draws are not independent.

```
# Using default of sample function, replace = FALSE
sample(x = Coin_box, size = 2)
## [1] "H" "T"
sample(x = Coin_box, size = 2)
## [1] "T" "H"
# following will give an error
# size of sample greater than tickets in box
#sample(x = Coin_box, size = 3)
```

If we replace the ticket, then the second draw is not affected by the first.

We can now explore what happens in a longer sequence of flips, where we replace the tickets after drawing, thus ensuring the draws are independent.

```
set.seed(30) # for reproducibility
sample(Coin_box, 6, replace = TRUE)
## [1] "T" "H" "T" "T" "H" "T"
```

We know, since we generated the sequence, that each draw is independent. However, if we only see the sequence, and don't know how it was generated, we may see a pattern, and doubt that it is random. However, the next sequence is different.

```
sample(Coin_box, 6, replace = TRUE)
## [1] "T" "H" "T" "H" "T" "H"
```

In his book, Thinking, Fast and Slow, Kahneman (2011, p.115) says that we have difficulty with truly random events. If we see what appears to be a regularity, we suspect the randomness of a process. In the statistical approach, we see what happened in the context of all the possible outcomes.

☞ **Your Turn**   Create Coin_box, and make 5 draws with replacement a few times.

9.2.2   Binomial Distribution

We now explore the binomial distribution: the structure is that of flipping a coin several times and summing the results. It is useful to denote the tails by 0 and the heads by 1. We draw 30 tickets from a box of 0s and 1s, Box01, and sum the values of the tickets.

```
Box01 <- c(0,1)
Samp <- sample(Box01,30, replace = TRUE)
Samp[1:10]
##  [1] 1 1 0 0 1 1 0 1 0 1
 table(Samp)
## Samp
##  0  1
## 17 13
 sum(Samp)
## [1] 13
```

We do this once more.

```
Samp <- sample(Box01,30, replace = TRUE)
 Samp[1:10]
##  [1] 0 1 0 1 0 0 0 1 0 0
 table(Samp)
```

```
## Samp
##  0  1
## 14 16
 sum(Samp)
## [1] 16
```

We have got a different number of 0s and 1s this time.

We will now take 30 draws from Box01 at a time, sum the values and repeat the process 10 times. We create a tibble of the 10 sums.

```
sims <- 10 # of simulations
sample_size <- 30 # sample size
sum_1 <- numeric(sims) # will store the sum in this vector
for (i in 1:sims) {
  Samp <- sample(Box01, sample_size,
                 replace = TRUE)
  # sample from Box01
  sum_1[i] <- sum(Samp) # Store the sum of values of each sample
}

sum_1 <- tibble(sum_1)
sum_1
## # A tibble: 10 x 1
##      sum_1
##      <dbl>
## 1      11
## 2      13
## 3      18
## 4      15
## 5      13
## 6      21
## 7      19
## 8      15
## 9      13
## 10     20
```

Sum of the values (see Fig. 9.1).

```
ggplot(sum_1, aes(x = sum_1)) +
  geom_bar() +
  xlim(5,25)
```

We will now take 30 draws from Box01 at a time, sum the values, but will repeat the process 1000 times.

Fig. 9.1 Sum of 1s from 30 draws from Box01, 10 repetitions

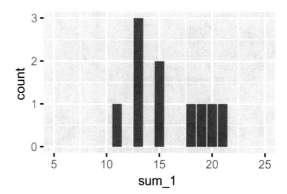

Fig. 9.2 Sum of 1s from 30 draws from Box01, 1000 repetitions

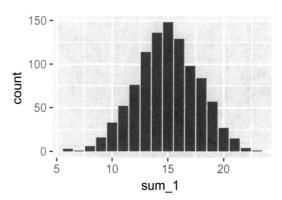

```
Box01 <- c(0,1)
sims <- 1000
sum_1 <- numeric(sims)
for (i in 1:sims) {
  Samp <- sample(Box01, 30, replace = TRUE)
  sum_1[i] <- sum(Samp)
}

sum_1 <- tibble(sum_1)

ggplot(sum_1, aes(x = sum_1)) +
  geom_bar()

# see Figure 9.2
```

The distribution of sums of the tickets looks approximately like a normal distribution (Fig. 9.2). We now plot a density curve of the distribution of sums and superimpose a normal distribution with the same mean and standard deviation.

Fig. 9.3 Dashed line: density curve of sum of 1s, as in Fig. 9.2. Solid line: density curve of normal distribution with same mean and standard deviation

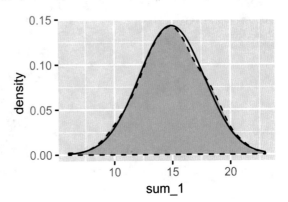

```
ggplot(sum_1, aes(x = sum_1)) +
  geom_density(fill = "grey80", linetype = 2) +
  # normal distribution
  stat_function(fun = dnorm, args =
          list(mean = mean(sum_1$sum_1),
               sd = sd(sum_1$sum_1)),
        linetype = 1)
```

The sum of the 1s from 30 independent coin tosses can be approximated by a normal distribution (Fig. 9.3).

9.2.3   Function for Binomial Distribution

We now create a function that we can use to explore the binomial distribution.

```
### Function for binomial distribution
sumBox01fun <- function(Box01 = c(0,1), sims = 1000, Size = 30)
{
sum_1 <- numeric(sims)
for (i in 1:sims) {
  Samp <- sample(Box01, Size, replace = TRUE)
  sum_1[i] <- sum(Samp)
}

sum_1 <- tibble(sum_1)

ggplot(sum_1, aes(x = sum_1)) +
  geom_bar()
}
### End of function
```

We can experiment with the function, for example:

```
sumBox01fun(Box01 = c(0,1,1), Size = 20, sims = 100)
```

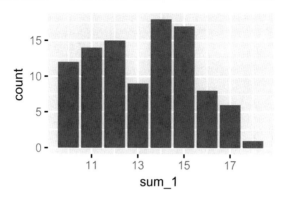

☞ **Your Turn**   Either write your own code or paste the code above and run
the function `sumBox01fun` providing inputs as required
(inputs are case sensitive).

1. Use Box = c(0,1,1,1), Size = 20 and sims = 1000 as inputs.
2. Use Box = c(0,1,1,1), Size = 200 and sims = 1000 as inputs.

9.3 Sampling Distribution

We will now examine the concept of sampling distributions, a concept that is central
to understanding econometrics (Kennedy 2003).

Let β^* be an estimator or a formula for a population parameter β. Then according
to Kennedy (2003, p. 404), 'Using β^* to produce an estimate of β can be conceptu-
alized as the econometrician shutting her or his eyes and obtaining an estimate of β
by reaching blindly into the sampling distribution of β^* to obtain a single number'.
We may want β^* to have good properties, for example, we want it to be unbiased.
β^* is unbiased if the mean of its sampling distribution equals β.

Figure 9.4 is a schematic of the process that leads us to a sampling distribution. If
we draw many samples from a population and calculate a statistic for each sample,
then the distribution of the statistic is the sampling distribution of the statistic.

Fig. 9.4 Sampling
distribution

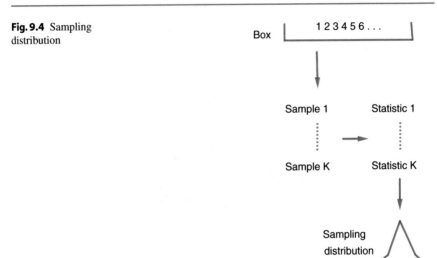

9.3.1   Six-Sided Dice Simulation

We consider a box that corresponds to six-sided dice; a box that has the tickets 1 to
6 in it, 10000 each. Because the population is very large relative to the size of the
samples that we will draw, each draw of a ticket can be considered independent of
another.

```
Box <- rep(1:6, 10000)
table(Box)
## Box
##     1     2     3     4     5     6
## 10000 10000 10000 10000 10000 10000
mean(Box)
## [1] 3.5
sd(Box)
## [1] 1.707839
```

We will draw 16 tickets at a time, and calculate the mean and standard deviation
of the tickets in the sample.

```
samp_size <- 16
Samp <- sample(Box, size = samp_size)
Samp
##  [1] 6 5 3 3 4 5 6 4 2 2 6 5 4 4 3 3
  sample_mean <- mean(Samp)
  sample_mean
## [1] 4.0625
  sample_sd <- sd(Samp)
  sample_sd
```

```
## [1] 1.340087
```

We repeat.

```
samp_size <- 16
Samp <- sample(Box, size = samp_size)
Samp
##  [1] 3 6 4 4 6 4 2 4 2 6 3 4 2 6 4 3
  sample_mean <- mean(Samp)
  sample_mean
## [1] 3.9375
  sample_sd <- sd(Samp)
  sample_sd
## [1] 1.436141
```

Each sample is different, with a different mean and standard deviation. We will now do a simulation, and draw samples of size 16 a 100 times. We use a for loop.

```
set.seed(34)
samp_size <- 16
simuls <- 100
sample_mean <- numeric(simuls)
sample_sd <- numeric(simuls)

for (i in 1:simuls) {
  Samp <- sample(Box, size = samp_size)
  sample_mean[i] <- mean(Samp)
  sample_sd[i] <- sd(Samp)
}

sample_mean_store <- sample_mean
```

We will now plot the distribution of the sample means (Fig. 9.5).

```
samp_dist <- tibble(sample_mean, sample_sd)
ggplot(samp_dist, aes(sample_mean)) +
  geom_histogram()

 paste("The mean of the sample means is:",
       sep = " ", round(mean(sample_mean),2))
## [1] "The mean of the sample means is: 3.46"
sd(sample_mean)
## [1] 0.4028505
sd(Box) / sqrt(samp_size)
## [1] 0.4269598
```

Fig. 9.5 Distribution of means of samples of size 16 taken from Box (equivalent to rolling of dice), with replacement

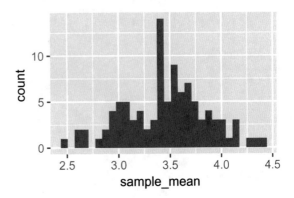

Fig. 9.6 Dashed line: density curve of means of samples as in previous figure, solid line: superimposed normal distribution

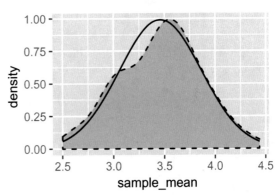

The sample mean is an unbiased estimator. The mean of the sample means to a close approximation equals the mean of the box. The standard error or the standard deviation of the distribution of the sample means is approximately equal to the standard deviation of the box divided by the square root of the size of a sample.

```
ggplot(samp_dist, aes(x = sample_mean)) +
  geom_density(fill = "grey80", linetype = 2) +
  stat_function(fun = dnorm, args =
        list(mean = mean(samp_dist$sample_mean),
             sd = sd(samp_dist$sample_mean)),
        linetype = 1)
```

The sampling distribution of the means approximately matches a normal distribution (Fig. 9.6).

We will now plot the distribution of the sample standard deviations (Fig. 9.7).

```
ggplot(samp_dist, aes(sample_sd)) +
  geom_histogram()
```

```
## `stat_bin()` using `bins = 30`. Pick better
## value with `binwidth`.
```

Fig. 9.7 Sampling
distribution of sample
standard deviation

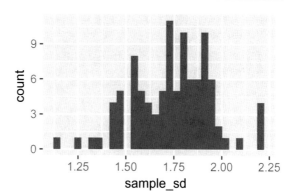

```
mean(sample_sd)
## [1] 1.741702
```

The mean of the distribution of the sample standard deviations is also approximately equal to the standard deviation of the box.

9.3.2   Function for Sampling Distribution

We put the code developed above into a function that will help us explore the sampling distribution of the mean.

```
### Function
samp_dist_mean_fun <- function( Box = rep(1:6, 10000),
  samp_size = 16, simuls = 100) {
sample_mean <- numeric(simuls)
sample_sd <- numeric(simuls)

for (i in 1:simuls) {
  Samp <- sample(Box, size = samp_size)
  sample_mean[i] <- mean(Samp)
  sample_sd[i] <- sd(Samp)
} # end of loop

samp_dist <- tibble(sample_mean, sample_sd)
gg1 <- ggplot(samp_dist, aes(sample_mean)) +
  geom_histogram(bins = 15)
print(gg1)
print(paste("The mean of the Box is:",
  sep = " ", round(mean(Box),2)))
print(paste("The standard deviation of the Box is:",
  sep = " ", round(sd(Box),2)))
print(paste("The mean of the sample means is:",
```

```
  sep = " ", round(mean(sample_mean),2)))
print(paste("The sd of the sample means is:", sep = " ",
  round(sd(sample_mean),2)))
 mean(sample_mean)
sd(sample_mean)
print(paste("sd(Box) / sqrt(samp_size) is:",
  #
sep = " ",
    round(sd(Box) / sqrt(samp_size),2)))
}
### end of function
```

☞ **Your Turn**   Either write your own code or paste the code above and run
the function `samp_dist_mean_fun` providing inputs
as required (inputs are case sensitive).

1. Use `Box = rep(1:6, 10000)`, `samp_size = 3`, `simuls = 1000`
   as inputs.
2. Use `Box = rep(c(1,1,1,2,2,6,9), 10000)`, `simuls = 1000` as
   inputs. Try with `samp_size = 3, 16, and 100`.

9.3.3   Sampling Distribution for the T-Statistic

We now calculate the t-statistic for each sample and plot the sampling distributions
for the t-statistics. The t-statistic is given by:
(mean of sample − mean of box)/standard error of mean

```
Box <- rep(1:6, 10000)
sd(Box)
## [1] 1.707839
mean(Box)
## [1] 3.5
```

The loop now contains a line to calculate the t-statistic.

```
set.seed(34)
samp_size <- 16
simuls <- 1000
se_samp <- numeric(simuls)
sample_tstat <- numeric(simuls)
for (i in 1:simuls){
```

Fig. 9.8 Sampling distribution of t-statistics

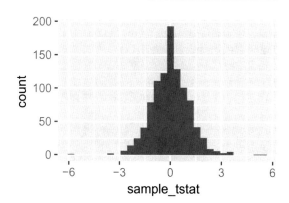

Fig. 9.9 Dashed line: density curve of t-statistics from samples; solid line: t-distribution

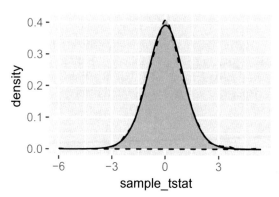

```
Samp <- sample(Box, size = samp_size)
se_samp <- sd(Samp)/sqrt(samp_size)
sample_tstat[i] <- (mean(Samp) -
    mean(Box))/se_samp # t statistic
}


tstat <- tibble(sample_tstat)
ggplot(tstat, aes(x = sample_tstat)) +
  geom_histogram()
```

Although the box is a uniform discrete distribution, and very different from a normal distribution, the distribution of t-statistics from samples taken from the box follows a t-distribution (Figs. 9.8, 9.9).

```
ggplot(tstat, aes(x = sample_tstat)) +
  geom_density(fill = "grey80", linetype = 2) +
stat_function(fun = dt, args =
    list(df = samp_size - 1),
    linetype = 1)
```

In his pioneering article on the t-distribution, Gossett had drawn samples of 4 from 3000 observations on the height of criminals and the length of their left middle finger (Bruce and Bruce 2017).

9.3.4 Inference from One Sample

In practice, we only have one sample of data. Having established the properties of the sampling distribution of the mean, we can now see how we carry out statistical inference if we have one sample. We are guided by the properties of the estimator (the formula used to estimate a statistic), as revealed by its sampling distribution.

Continuing from above, we once again take a sample from Box.

```
Samp <- sample(Box, size = samp_size)
```

Having drawn the sample, we estimate the mean by calling the mean() function, and by using the lm() function. Both give us the same estimate of the mean, but lm() also gives us other statistics. Plus lm() generalizes to linear regression with a dependent variable and several independent variables.

```
mean(Samp)
## [1] 4.4375
mod_Samp <- lm(Samp ~ 1)
summary(mod_Samp)
##
## Call:
## lm(formula = Samp ~ 1)
##
## Residuals:
##    Min    1Q Median    3Q    Max
## -3.438 -1.438  1.062  1.562  1.562
##
## Coefficients:
##             Estimate Std. Error t value Pr(>|t|)
## (Intercept)   4.4375     0.4741    9.36 1.18e-07
##
## (Intercept) ***
## ---
## Signif. codes:
## 0 '***' 0.001 '**' 0.01 '*' 0.05 '.' 0.1 ' ' 1
##
## Residual standard error: 1.896 on 15 degrees of freedom
```

Fig. 9.10 Distribution of
t-statistics from samples of
Box, and the t-statistic
calculated from one sample
(vertical dashed line), under
the null hypothesis that mean
of Box = 0

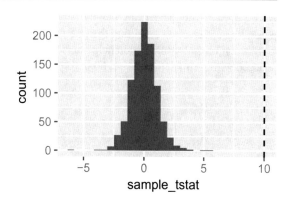

The summary shows the estimate of the mean. The t-statistic is very high. We know it is very high because we had plotted the t-distribution that we got from Box (Fig. 9.8); now we also plot the value of the t-statistic in the sample to the t-distribution. Figure 9.10 shows us that the probability of observing such a high, or higher, t-statistic if the null hypothesis of the mean being zero were true, is very small.

According to Maindonald and Braun (2010, p. 108), 'The formal methodology of hypothesis testing may seem contorted. A small p-value makes the null hypothesis appear implausible. It is not a probability statement about the null hypothesis itself, or for that matter about its alternative. All it offers is an assessment of implications that flow from accepting the null hypothesis'.

```
ggplot(tstat, aes(x = sample_tstat)) +
  geom_histogram() +
  geom_vline(xintercept = 10, linetype = 2)
```

We can get the confidence interval for the mean.

```
confint(mod_Samp)
##                 2.5 %  97.5 %
## (Intercept) 3.42705 5.44795
```

We now turn to a graphical understanding of confidence intervals.

9.3.5 Confidence Intervals

Confidence intervals indicate the estimate and also the uncertainty of the estimate. We often use 95 percent confidence intervals. A particular confidence interval may or may not contain the mean. Using C to denote the confidence level, Moore et al. (2009, p. 359) define a confidence interval: 'A level C confidence interval for a parameter is an interval computed from sample data by a method that has probability C of producing an interval containing the true value of the parameter'.

Simulation and visualization help convey the concept of confidence intervals. We proceed with a simulation as we did before.

```
Box <- rep(1:6, 10000)
sd(Box)
## [1] 1.707839
mean(Box)
## [1] 3.5
```

Now while carrying out the simulation we will calculate the confidence intervals.

```
Samp <- sample(Box, size = samp_size)
Conf <- confint(lm(Samp ~ 1), level = 0.95)
Conf
##                     2.5 %   97.5 %
## (Intercept) 2.71507 4.40993
  Conf[1]
## [1] 2.71507
  Conf[2]
## [1] 4.40993
```

We now do this a number of times.

```
set.seed(34)
samp_size <- 16
simuls <- 100
sample_mean <- numeric(simuls)
conf_lo <- numeric(simuls)
conf_hi <- numeric(simuls)
for (i in 1:simuls){
  Samp <- sample(Box, size = samp_size)
  sample_mean[i] <- mean(Samp)

  Conf <- confint(lm(Samp ~ 1), level = 0.95)
  conf_lo[i] <- Conf[1]
  conf_hi[i] <- Conf[2]
  }
```

Having simulated the confidence intervals, we now examine them.

```
samp_index <- 1:simuls
Conf_int <- tibble(conf_lo, conf_hi,
          samp_index, sample_mean)

head(Conf_int,20)
```

```
## # A tibble: 20 x 4
##    conf_lo conf_hi samp_index sample_mean
##      <dbl>   <dbl>      <int>        <dbl>
##  1    1.71    3.54          1         2.62
##  2    2.45    4.43          2         3.44
##  3    2.04    4.08          3         3.06
##  4    2.53    4.22          4         3.38
##  5    2.80    4.83          5         3.81
##  6    2.31    4.44          6         3.38
##  7    2.27    3.85          7         3.06
##  8    3.14    5.11          8         4.12
##  9    3.04    4.84          9         3.94
## 10    2.80    4.45         10         3.62
## 11    2.47    4.53         11          3.5
## 12    2.76    4.61         12         3.69
## 13    1.94    3.94         13         2.94
## 14    2.13    3.87         14            3
## 15    2.25    4.62         15         3.44
## 16    3.28    4.97         16         4.12
## 17    2.19    4.06         17         3.12
## 18    2.59    4.53         18         3.56
## 19    2.40    4.73         19         3.56
## 20    2.90    4.73         20         3.81
Conf_int_20 <- Conf_int[1:20,]

Covers <- 1 - ifelse(conf_lo > 3.5 |
                      conf_hi < 3.5, 1, 0)
sum(Covers)/simuls
## [1] 0.97
```

Above we have calculated the percentage of confidence intervals that cover the mean.

```
ggplot(Conf_int_20) +
  geom_pointrange(aes(x = samp_index,
                      y = sample_mean,
                      ymin = conf_lo,
                      ymax = conf_hi)) +
  geom_hline(yintercept = 3.5, linetype = 2)
```

The first 20 confidence intervals are plotted in Fig. 9.11. The confidence intervals vary in their width.

Fig. 9.11 20 simulated confidence intervals of mean of box (six-sided dice); true mean is 3.5

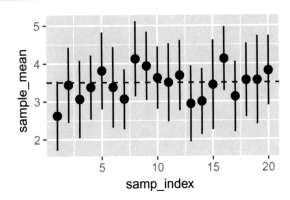

Fig. 9.12 Bootstrap process

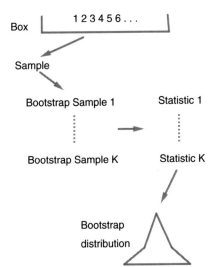

9.4   Bootstrap

The bootstrap is a way of estimating the uncertainty in a test statistic that relies on the sample itself and uses the computer for repeated calculations.

While investigating the sampling distribution via simulation we drew samples repeatedly from the Box. Unfortunately, we don't get to see the Box, only a sample from a Box. We can pull ourselves up by our bootstraps if we repeatedly sample from the sample. If the sample is like the population, then many samples from the sample are representative of many samples from the population (Chihara and Hesterberg 2011).

Figure 9.12 is a schematic diagram of the bootstrap process. In contrast with sampling distributions, we take bootstrap samples from the sample itself, by sampling with replacement. For each bootstrap sample, we calculate a test statistic.

We take bootstrap samples by sampling with replacement.

```
Box <- rep(1:6, 10000)
sd(Box)
## [1] 1.707839
mean(Box)
## [1] 3.5
```

We pull out a sample from the box–`BoxSample1`.

```
samp_size <- 100
Box_Sample1 <- sample(Box, size = samp_size)
Box_Sample1[1:10]
##  [1] 4 5 3 1 6 5 3 4 1 4
  mean(Box_Sample1); sd(Box_Sample1)
## [1] 3.28
## [1] 1.670178
```

The sampling distribution is the distribution of sample means where each sample is taken from the Box. The bootstrap distribution is the distribution of sample means where we sample from the sample of data that we have; here the sample of data is `Box_Sample1`. We will compare the bootstrap distribution with the sampling distribution.

```
set.seed(34)
simuls <- 10000
Sample_mean <- numeric(simuls)
Bootstrap_mean <- numeric(simuls)

for (i in 1:simuls){
  # sample from the sample
  Boot_sample <- sample(Box_Sample1,
            size = samp_size, replace = TRUE)
  Bootstrap_mean[i] <- mean(Boot_sample)
  # sample from the Box
  Sample <- sample(Box, size = samp_size)
  Sample_mean[i] <- mean(Sample)
}
```

We examine different aspects of the bootstrap and sampling distributions.

```
Distribution <- c(Bootstrap_mean, Sample_mean)

Type <- c(rep("Bootstrap", simuls),
            rep("Sampling", simuls))

str(Distribution)
## num [1:20000] 3.47 3.44 3.53 3.46 3.41 3.5 3.37 3.5 3.27
```

Fig. 9.13 Bootstrap and
sampling distribution

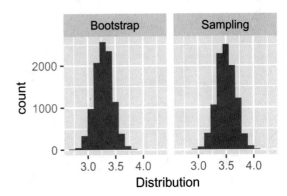

```
                      3.24 ...
str(Type)
##   chr [1:20000] "Bootstrap" "Bootstrap" ...
Boot_sim <- tibble(Distribution, Type)
Boot_sim[5,]
## # A tibble: 1 x 2
##     Distribution Type
##            <dbl> <chr>
## 1           3.41 Bootstrap
```

```
ggplot(Boot_sim, aes(x = Distribution)) +
  geom_histogram(bins = 15) +
  facet_wrap(Type ~ .)
```

```
Boot_sim %>%
  group_by(Type) %>%
  summarize(sd = sd(Distribution),
            mean = mean(Distribution))
## # A tibble: 2 x 3
##    Type          sd   mean
##    <chr>      <dbl>  <dbl>
## 1 Bootstrap 0.166   3.28
## 2 Sampling  0.170   3.50
```

The shapes of the bootstrap and sampling distributions are similar (Fig. 9.13). The standard deviation of the bootstrap is close to that of the sampling distribution. The mean of the bootstrap distribution is equal to the mean of the sample from which the bootstrap samples are drawn.

One method of estimating a confidence interval is to use the quantiles of the bootstrap distribution–the bootstrap percentile method.

```
round(quantile(Bootstrap_mean, probs = c(0.025, 0.975)),2)
##  2.5% 97.5%
##  2.96  3.61
```

We have used the bootstrap percentile method, which works well in this case, and is intuitive. There are several refinements of the bootstrap. Maindonald and Braun point out (2010, p. 132), 'Bootstrap methods are not a panacea. We must respect the structure of the data; any form of dependence in the data must be taken into account. There are contexts where the bootstrap is invalid and will mislead. As a rough guideline, the bootstrap is unlikely to be satisfactory for statistics, including maximum, minimum and range, that are functions of sample extremes. The bootstrap is usually appropriate for statistics from regression analysis–means, variances, coefficient estimates, and correlation coefficients'.

9.4.1 Function to Understand Bootstrap

We use the code developed above to construct a function that helps us understand the bootstrap.

```
### Function to understand bootstrap

Boot_understand_fun <- function(Box = rep(1:6, 10000),
  Seed = 34, simuls = 1000, samp_size = 100) {
set.seed(Seed)
Box_Sample1 <- sample(Box, size = samp_size)
Sample_mean <- numeric(simuls)
Bootstrap_mean <- numeric(simuls)

# loop
for (i in 1:simuls) {
  Boot_sample <- sample(Box_Sample1,
        size = samp_size, replace = TRUE)
  Bootstrap_mean[i] <- mean(Boot_sample)
  Sample <- sample(Box, size = samp_size)
  Sample_mean[i] <- mean(Sample)
}

# end loop

Distribution <- c(Bootstrap_mean, Sample_mean)

Type <- c(rep("Bootstrap", simuls),
            rep("Sampling", simuls))
```

```
Boot_sim <- tibble(Distribution, Type)
Boot_sim

gg_boot <- ggplot(Boot_sim, aes(x = Distribution)) +
  geom_histogram(bins = 15) +
  facet_wrap(Type ~ .)

print(gg_boot)


print(paste("The mean of the box =", sep = " ", mean(Box)))
print(paste("The sd of the box =", sep = " ", round(sd(Box),2)))

Boot_sim_stats <- Boot_sim %>%
  group_by(Type) %>%
  summarize(sd = round(sd(Distribution),3),
            mean = round(mean(Distribution),2))
print(Boot_sim_stats)

print(paste("The quantiles of the"))
print(paste("sampling distribution of the mean are:"))
print(quantile(Sample_mean, probs = c(0.025, 0.975)))
print(paste("The quantiles of the
  bootstrap distribution of the mean are:"))
print(quantile(Bootstrap_mean, probs = c(0.025, 0.975)))


}

### end function

Boot_understand_fun()

## [1] "The mean of the box = 3.5"
## [1] "The sd of the box = 1.71"
## # A tibble: 2 x 3
##    Type         sd   mean
##    <chr>      <dbl> <dbl>
## 1 Bootstrap  0.18   3.29
## 2 Sampling   0.17   3.5
## [1] "The quantiles of the"
## [1] "sampling distribution of the mean are:"
##    2.5%    97.5%
## 3.18975 3.84000
## [1] "The quantiles of the \n  bootstrap distribution of the
       mean are:"
```

```
##      2.5%    97.5%
## 2.93975 3.65000
```

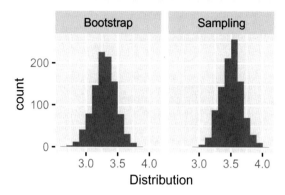

☞ **Your Turn**   Use `Boot_understand_fun()`, supplying your own inputs to compare the bootstrap and sampling distributions.

9.5   Permutation Tests

We discuss permutation tests; these tests are also called randomization tests. We use a hypothetical example to understand permutation tests. We have scores on a test in class A and B; both have five students. We input the data.

```
scores <- c(10, 11, 12, 13, 14, 15, 16, 17, 18, 19)
student <- c("Student1", "Student2",
   "Student3", "Student4", "Student5",
   "Student6", "Student7","Student8",
   "Student9", "Student10")
Class <- c("A","A", "A", "A","A","B","B","B", "B","B")
```

The scores are as follows:

```
Scores <- tibble(scores, student, Class)
Scores
## # A tibble: 10 x 3
##     scores student   Class
##      <dbl> <chr>     <chr>
## 1      10 Student1  A
## 2      11 Student2  A
## 3      12 Student3  A
## 4      13 Student4  A
```

```
## 5      14 Student5  A
## 6      15 Student6  B
## 7      16 Student7  B
## 8      17 Student8  B
## 9      18 Student9  B
## 10     19 Student10 B
```

We can estimate the mean score in class B and the mean score in class A and then take the difference: the observed mean difference is 5:

```
lm(scores ~ Class)
##
## Call:
## lm(formula = scores ~ Class)
##
## Coefficients:
## (Intercept)        ClassB
##          12             5
```

So the students in Class B performed better on average. Could this be due to chance? To test whether the difference in scores could reflect chance or not, we take our null hypothesis to be: Class B and Class A scores are identically distributed. In a world where the null hypothesis is true, the students would perform the same whether they are in Class A or Class B. So student1 would get a score of 10 whether the student is in Class A or Class B.

In terms of drawing from a box, we say that scores observed in classes A and B both belong to the same box. We take repeated draws of size ten, without replacement from this box. The first five tickets drawn are labelled as class A and the next five as class B. We then take the difference of means.

The idea is to shuffle the tickets, sort them into Class A and Class B, and take the difference of means.

Figure 9.14 is a schematic of the permutation process.

We now implement this in R. We first pick five of the ten students randomly and put them in class A.

```
index_Ap <- sample(1:10,5); index_Ap
## [1] 2 1 5 6 8
Scores[index_Ap,]
## # A tibble: 5 x 3
##    scores student  Class
##     <dbl> <chr>    <chr>
## 1      11 Student2  A
## 2      10 Student1  A
## 3      14 Student5  A
## 4      15 Student6  B
## 5      17 Student8  B
```

Fig. 9.14 Permutation
process

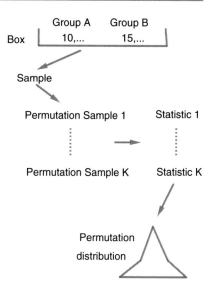

So student4 who scored 13, and was in Class A, is now labelled as Class A.
Student10 who is actually in Class B, is labelled as Class A, since under the null
hypothesis the score would be the same whether student10 is in Class A or Class B.
We then put the remaining five students in class B.

```
Scores[-index_Ap,]
## # A tibble: 5 x 3
##    scores student   Class
##     <dbl> <chr>     <chr>
## 1      12 Student3  A
## 2      13 Student4  A
## 3      16 Student7  B
## 4      18 Student9  B
## 5      19 Student10 B
```

We then take the difference of means.

```
mean(scores[index_Ap]) -
   mean(scores[-index_Ap])
## [1] -2.2
```

This shuffling is repeated with a loop.

```
iters <- 999
mean_diff_P <- numeric(iters)
for (i in 1:iters) {
  index_Ap <- sample(1:10,5); index_Ap
mean_diff_P[i] <-   mean(scores[index_Ap]) -
   mean(scores[-index_Ap])
}
```

Fig. 9.15 Permutation distribution of mean differences. Observed value = 5

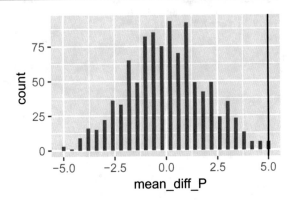

We collect the values in a tibble.

```
mean_diff <- tibble(mean_diff_P)
mean_diff
## # A tibble: 999 x 1
##     mean_diff_P
##           <dbl>
## 1      -0.200
## 2      -0.6
## 3      -1
## 4      -0.6
## 5      -1.8
## 6       3
## 7       2.20
## 8       0.6
## 9      -4.2
## 10      1.8
## # ... with 989 more rows
```

We now plot the permutation distribution of mean differences (Fig. 9.15).

```
ggplot(mean_diff, aes(x = mean_diff_P)) +
  geom_bar() +
  geom_vline(xintercept = 5)
```

We count the number of times where the absolute value of the mean difference is equal to, or greater than, 5.

```
sum_extreme_values <- sum(abs(mean_diff_P )>=5)
sum_extreme_values
## [1] 9
```

We calculate the p-value. We add 1 to numerator and denominator for an improved p-value estimate (Chihara and Hesterberg 2011).

```
p_value <- (sum_extreme_values + 1) / (iters + 1)
paste("The p-value is", sep = " ", round(p_value, 4))
## [1] "The p-value is 0.01"
```

Our alternate hypothesis is that Class B students are better. The probability of observing a mean difference of 5 if the sharp null hypothesis that the scores of all the students are the same is true is very low; the data provides strong evidence against the sharp null hypothesis.

9.6   Example: Verizon

We study a case presented by Chihara and Hesterberg (2011) involving a phone company, Verizon. By regulation, Verizon was responsible for carrying out repairs not only for its own customers (ILEC) but also for the customers of competing companies (CLEC). The New York Public Utilities Commission monitored whether Verizon was being equally fair to own and competitor's customers. Many hypothesis tests were carried out, and if more than 1 percent of the tests were significant, then Verizon would pay large penalties. The Verizon data is available in the resample package.

```
library(resample)
```

```
## Registered S3 method overwritten by 'resample':
##   method              from
##   print.resample modelr
```

```
data(Verizon)
str(Verizon)
## 'data.frame':    1687 obs. of  2 variables:
##  $ Time : num  17.5 2.4 0 0.65 22.23 ...
##  $ Group: Factor w/ 2 levels "CLEC","ILEC": 2 2 2 2 2 2 2
     2 2 2 ...
table(Verizon$Group)
##
## CLEC ILEC
##   23 1664
```

```
ggplot(Verizon, aes(x = Time)) +
  geom_density() +
  facet_wrap(~ Group)
```

The density curves of repair times are very skewed (Fig. 9.16).

We use a log transformation of the distributions and then make boxplots. The median of the CLEC repair times is higher than the ILEC repair times (Fig. 9.17).

Fig. 9.16 Density curves for
repair times of CLEC and
ILEC customers

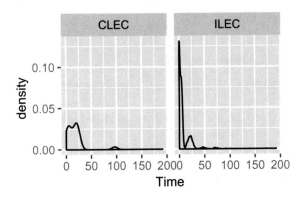

Fig. 9.17 Boxplots of CLEC
and ILEC customer repair
times

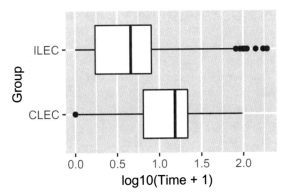

```
ggplot(Verizon, aes(x = Group,
                    y = log10(Time + 1))) +
  geom_boxplot() +
  coord_flip()
```

```
Verizon %>%
  group_by(Group) %>%
  summarize(mean_T = round(mean(Time),1),
    t_mean_T = round(mean(Time, trim = 0.25),1))
## # A tibble: 2 x 3
##    Group mean_T t_mean_T
##    <fct>  <dbl>    <dbl>
## 1 CLEC    16.5     13.8
## 2 ILEC     8.4      3.5
```

The mean repair time and the trimmed mean repair time are higher for CLEC
customers in this sample.

```
Ver_CLEC <- Verizon %>%
  filter(Group == "CLEC")
Ver_ILEC <- Verizon %>%
  filter(Group == "ILEC")
```

The trimmed mean for CLEC customers is about 10 more than for ILEC customers.

```
diff_t_mean <- mean(Ver_ILEC$Time, trim = .25) -
  mean(Ver_CLEC$Time, trim = .25)
diff_t_mean
## [1] -10.336
```

9.6.1   Permutation Test

We now carry out a randomization test of the difference of trimmed means. We take the Verizon times of both CLEC and ILEC customers. The null hypothesis is that the repair times are the same whether the customer is a CLEC or ILEC customer. We repeatedly sample from the Verizon repair times and randomly allocate 23 scores to CLEC and 1664 scores to ILEC and then take the difference of means.

```
iters <- 1000
diff_t_perms <- numeric(iters)
for(i in 1:iters) {
  index <- sample(1:1687, size=1664, replace = FALSE)
  diff_t_perms[i] <- mean(Verizon$Time[index], trim = .25) -
    mean(Verizon$Time[-index], trim = .25)
}
```

We plot the permutation distribution for the difference of trimmed means (Fig. 9.18).

```
ggplot() +
  geom_histogram(aes(diff_t_perms))  +
  geom_vline(xintercept = diff_t_mean,
             linetype=2)

#P-value difference in trimmed means
p_value <- (sum(diff_t_perms <=
       diff_t_mean) + 1)/(iters+ 1)
paste("The p_value is", sep = " ", round(p_value,4))
## [1] "The p_value is 0.002"
```

We see that the p-value is very low.

Fig. 9.18 Permutation distribution of difference of trimmed means (ILEC - CLEC)

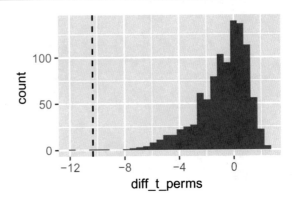

9.6.2   Bootstrapping Confidence Intervals

The CLEC and ILEC data are skewed. In this context, the trimmed mean is a good estimator since it trims away the extreme observations.

One of the advantages of the bootstrap is that it can help us estimate the confidence intervals related to the difference of trimmed means. The bootstrap was devised for estimating statistical accuracy of estimators such as trimmed means, where there is no neat algebraic formula (Efron and Tibshirani 1991).

We had already filtered the data and have separate data for the CLEC customers, Ver_CLEC and for ILEC customers, Ver_ILEC. We will sample with replacement from these two vectors, take the mean of each and then the difference. We estimate the lengths of the customer repair time variable in each of these.

```
CLEC <- Ver_CLEC$Time
ILEC <- Ver_ILEC$Time
length_C <- length(CLEC)
length_C
## [1] 23
length_I <- length(ILEC)
length_I
## [1] 1664
```

We calculate the difference in repair times between the two types of customers observed in our sample.

```
mean(ILEC) - mean(CLEC)
## [1] -8.09752
mean(ILEC, trim = 0.25) - mean(CLEC, trim = 0.25)
## [1] -10.336
```

We will now do sampling with replacement from the data a number of times (Figs. 9.19 and 9.20).

Fig. 9.19 Histogram of bootstrap distribution of difference of means

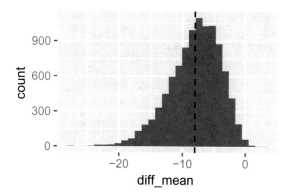

Fig. 9.20 Histogram of bootstrap distribution of difference in trimmed means

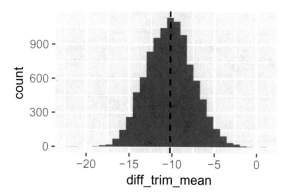

```
  sims <- 10000
diff_mean <- numeric(sims)
diff_trim_mean <- numeric(sims)

for (i in 1:sims){
  samp_I <- sample(ILEC,length_I,replace = TRUE)
  samp_C <- sample(CLEC,length_C,replace = TRUE)
  diff_mean[i] <- mean(samp_I) - mean(samp_C)
  diff_trim_mean[i] <- mean(samp_I, trim = 0.25) -
    mean(samp_C, trim = 0.25)
}


ggplot() + geom_histogram(aes(diff_mean)) +
  geom_vline(xintercept = mean(diff_mean),
             linetype = 2)


ggplot() +
  geom_histogram(aes(diff_trim_mean)) +
  geom_vline(xintercept = mean(diff_trim_mean), linetype = 2)
```

We see that the bootstrap distribution of the difference of means is skewed. There-fore, we prefer to use the trimmed mean estimates. We can calculate the bootstrap percentile interval below.

```
quantile(diff_trim_mean, probs = c(0.025, 0.975))
##        2.5%        97.5%
## -15.379100  -4.792572
```

9.7   Cautionary Example with Synthetic Data

Andrew Gelman has drawn attention to the perils of statistical inference when we have noisy data with a weak signal. If we use noisy data with a weak signal and a small sample size, statistical inference cannot separate the signal from the noise. It can be a mistake to think that a statistically significant result that has come through despite small sample size is the true signal—it may be very misleading.

In Gelman's words: 'we've seen from statistical analysis that the "What does not kill my statistical significance makes it stronger" attitude is a fallacy: Actually, the noisier the study, the *less* we learn from statistical significance'.

https://statmodeling.stat.columbia.edu/2017/02/06/

(accessed on 30 June 2019)

In the code below, xvar is a variable that is normally distributed with mean = 0.1 and sd = 1. What we observe is xobs which is xvar + noise. The noise is 2 * draws from a normal distribution with mean = 0 and sd = 1.

```
set.seed(15)
sample_size <- 30
xvar <- rnorm(sample_size, mean = 0.1, sd =1)
noise <- 2 * rnorm(sample_size)
xobs <- xvar + noise
summary(lm(xobs ~ 1))
##
## Call:
## lm(formula = xobs ~ 1)
##
## Residuals:
##     Min      1Q  Median      3Q     Max
## -4.6369 -1.5422  0.2838  1.6177  4.2065
##
## Coefficients:
##             Estimate Std. Error t value Pr(>|t|)
## (Intercept)   1.0294     0.3839   2.681    0.012
##
## (Intercept) *
## ---
```

```
## Signif. codes:
## 0 '***' 0.001 '**' 0.01 '*' 0.05 '.' 0.1 ' ' 1
##
## Residual standard error: 2.103 on 29 degrees of freedom
confint(lm(xobs ~ 1))
##                 2.5 %    97.5 %
## (Intercept) 0.2442544 1.814643
```

The estimate of the mean of xvar using the observed data xobs is 1.03, an order of magnitude greater than the true value, and which is statistically significant.

☞ **Your Turn**   Try out the code above with a different seed, and then with a larger sample size of 900.

9.8   Resources

For Better Understanding

Freedman et al. (2009) is a fantastic book, informal yet conceptually sophisticated and accurate. Bruce and Bruce (2017) is accessible and written for data scientists. Kennedy (2003) has a useful appendix on sampling distributions.

For Going Further

Chihara and Hesterberg (2011) use resampling and R.

References

Bruce, P., and A. Bruce. 2017. *Practical statistics for data scientists: 50 essential concepts.* Sebastopol: O'Reilly Media.

Chihara, L., and T. Hesterberg. 2011. *Mathematical statistics with resampling and R.* Hoboken, New Jersey: Wiley.

Efron, B., and R.J. Tibshirani. 1991. Statistical data analysis in the computer age. *Science, New Series* 253 (5018): 390–395.

Freedman, D., R. Pisani, and R. Purves. 2009. *Statistics,* 4th ed. New Delhi: Viva Books.

Kahneman, D. 2011. *Thinking, fast and slow.* India: Penguin Books.

Kennedy, P. 2003. *A guide to econometrics.* Cambridge: The MIT Press.

Maindonald, J., and W.J. Braun. 2010. *Data analysis and graphics using R: an example-based approach*, 3rd ed., Cambridge series in statistical and probabilistic mathematics Cambridge: Cambridge University Press.

Moore, D.S., G.P. McCabe, and B.A. Craig. 2009. *Introduction to the practice of statistics.* New York: W H Freeman and Company.

Causal Inference

10

10.1 Introduction

We may be interested in causal and non-causal questions:

Descriptive how does income vary across occupations?
Forecasting what will the price of crude oil be next year?
Causal if we change the ratio of students to teachers, will learning scores improve?

If we are interested in the answer to causal questions, we would want to use causal inference. Books in statistics and in econometrics often vary in how explicit they are in their treatment of causality. Causal inference is of great relevance in programme evaluation, a domain which, according to Abadie and Cattaneo (2018, p. 466), is 'expanding the social, biomedical, and behavioral sciences that studies the effect of policy interventions. The policies of interest are often governmental programs, like active labor market interventions or antipoverty programs'.

Morgan and Winship's (2014) wonderful book *Counterfactuals and Causal Inference* has the following quote from Gary King on the back cover: 'More has been learned about causal inference in the last few decades than the sum total of everything that had been learned about it in all prior recorded history'. Morgan and Winship's book makes use of two approaches to causal inference: (1) potential outcomes, or counterfactuals, and (2) causal graphs. They argue that these are complementary, and use both, as do authors like Abadie and Catteneo (2018); and that is the view we take in this chapter.

10.2   Causal Graphs and Potential Outcomes

10.2.1 Simple Example with Synthetic Data

To illustrate the causal graphs and potential outcomes approaches, we consider a case of some individuals who are affected with pain in their back. They can take a medicine to relieve their pain.

We denote pain by P and medicine by M, and the individuals are indexed by i. We want to examine the effect of M on P. M is a cause: we want to intervene to reduce pain. But is it effective? We can randomly assign M to different individuals.

We will use the structural causal model of Pearl et al. (2016) as a framework to understand the causal inference regarding M. The structural causal model consists of a causal graph and accompanying structural equations that show our assumptions about the phenomenon. Here the causal graph is $M \rightarrow P$, and the structural equation is (we assume linearity to simplify; Pearl's approach does not require the assumption of linearity) $P = \beta_0 + \beta_M M + U_P$ (see Fig. 10.1).

Since M is randomly assigned, it is independent of U_P. U_P captures the other causes or sources of variation in P.

In the last chapter we had used box models for random variables. The structural equation in this case has a box for M, and a box for U_P (Fig. 10.2).

We will now carry out a simulation. We assume that $\beta_0 = 10$, $\beta_M = -3$, and $U_P \sim N(0, 1)$.

```
library(tidyverse)
set.seed(22)
beta0 <- 10
betaM <- -3
num <- 400
U_P <- rnorm(num)
```

We generate the dataset `Meds`, which contains variables `M, P, U_P`.

```
set.seed(3)
M <- sample(c(0,1), num, replace = T)
Indiv <- 1:num
P <- beta0 + betaM * M + U_P
Meds <- tibble(Indiv, M, U_P, P)
kable(round(head(Meds),2))
```

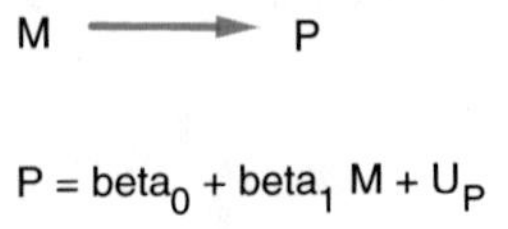

$$M \longrightarrow P$$

$$P = beta_0 + beta_1\, M + U_P$$

Fig. 10.1 Structural causal model: causal graph and accompanying structural equation

$$P = beta_0 + beta_1\, \boxed{M} + \boxed{U_P}$$

Fig. 10.2 Structural equation: Nature listens to the value of M and Up while fixing the value of P. M and Up are random variables, and therefore can be represented as boxes

Indiv	M	U_P	P
1	0	−0.51	9.49
2	1	2.49	9.49
3	1	1.01	8.01
4	0	0.29	10.29
5	1	−0.21	6.79
6	1	1.86	8.86

We see in the generated data that individual 1 is assigned to $M = 0$, i.e. is in the control group. We observe that the value of P for person one, when assigned $M = 0$ is 9.49. In the case of individual 2, the value of P for person two, when assigned $M = 1$ is 9.49.

We denote the potential outcome for an individual in this case by $P_i(m)$ where i indexes the individual and m is either 0 or 1, depending on the assigned value of M. Figure 10.3 shows how the structural causal graph is related to potential outcomes.

Here, $P_1(0) = 9.49$ and $P_2(1) = 9.49$.

The causal effect for an individual is: $P_i(1) - P_i(0)$.

Unfortunately, and this is called the fundamental dilemma of causal inference, we do not observe both potential outcomes for an individual. So we do not observe in this case the value of pain for person 1 if she had got the medicine, and the pain person 2 would have experienced had she not received the medicine. We do not observe the counterfactual.

Individual	M	$P_i(0)$	$P_i(1)$
1	0	9.49	?
2	1	?	9.49

We have to settle for an average treatment effect, the difference between the pain experienced by the control group and the treatment group, which we estimate below. We run a regression of P on M: $P = r_0 + r_1 M + e$, being the regression equation. The regression equation is distinct from the structural equation. In this case, r_1 will give us an estimate that is close to the true value, or the structural parameter, β_M; it is not necessarily the case that a regression coefficient will give us a good estimate of the causal effect.

```
library(texreg)
modd <- lm(P ~ M, data = Meds)
texreg(list(modd), caption = "Regression of P on M",
       caption.above = TRUE,
       include.adjrs = FALSE,
       include.rmse = FALSE)
```

$$M \mid m \longrightarrow P(m)$$

Fig. 10.3 Counterfactuals and potential outcomes. When the value of M is set to m, P has a potential outcome denoted by P(m)

Table 10.1 Regression of P on M

	Model 1
(Intercept)	9.99***
	(0.07)
M	−3.10***
	(0.10)
R^2	0.71
Num. obs.	400

$^*p < 0.05$; $^{**}p < 0.01$; $^{***}p < 0.001$

The estimated coefficient (-3.1) is close to the value of β_M (-3) (Table 10.1).

☞ **Your Turn**   Try simulating the data with a different value for the treatment effect β_M, and change the value in the function `set.seed()`.

10.2.2 Randomized Assignment of Treatment (Causal Graphs)

In the absence of randomized assignment of treatment, the level of medicine and pain may be influenced by a confounding variable (Fig. 10.4). Randomized assignment attempts to ensure that the treatment and control groups only differ with respect to the treatment, both with respect to observed and unobserved variables (Fig. 10.5). Once R (random assignment) determines the value of M, or M is set by R, the link between C and M is broken. C no longer acts as a confounder.

Fig. 10.4 In the absence of random assignment of M, a variable C may be a common cause of M and P, and will be a confounder variable

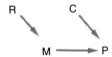

Fig. 10.5 Once R (random assignment) determines the value of M, or M is set by R, the link between C and M is broken. C no longer acts as a confounder

Table 10.2 Patient potential outcomes in Rubin's perfect doctor example

	Y_po1	Y_po0	Eff
1	14.00	13.00	1.00
2	0.00	6.00	−6.00
3	1.00	4.00	−3.00
4	2.00	5.00	−3.00
5	3.00	6.00	−3.00
6	1.00	6.00	−5.00
7	10.00	8.00	2.00
8	9.00	8.00	1.00

10.2.3 Randomized Assignment of Treatment (Potential Outcomes)

We consider an intriguing example provided by Rubin (2008). A treatment, surgery, affects years lived. Let $Y(0)$ denote the potential outcome without surgery, and $Y(1)$ denote the potential outcome with surgery. For coding in R we use `Y_po0`, and `Y_po1`, respectively.
We input the data.

```
Y_po0 <- c(13,6,4,5,6,6,8,8)
Y_po0
## [1] 13  6  4  5  6  6  8  8
Y_po1 <- c(14,0,1,2,3,1,10,9)
Y_po1
## [1] 14  0  1  2  3  1 10  9


Eff <- Y_po1 - Y_po0
```

We create a data frame.

```
surg <- data.frame(Y_po1, Y_po0, Eff)
```

We print the data for the perfect doctor example (Table 10.2).

```
library(xtable)
sable <- xtable(surg, caption = "Patient potential outcomes in
  Rubin's perfect doctor example")
print(sable, caption.placement = "top")
```

In this example, the effects vary a lot between the 8 people who are candidates for surgery (Table 10.2).

```
library(skimr)
skim1 <- surg %>%
  skim_to_wide()
skimble <- xtable(skim1[,c(2,6)], caption = "Means")
print(skimble, caption.placement = "top")
```

Table 10.3 Means

	Variable	Mean
1	Eff	−2
2	Y_po0	7
3	Y_po1	5

Table 10.4 Treatment assignment and observed outcomes

	D	Yi
1	1.00	14.00
2	1.00	0.00
3	1.00	1.00
4	1.00	2.00
5	0.00	6.00
6	0.00	6.00
7	0.00	8.00
8	0.00	8.00

The true average causal effect is −2 (Table 10.3). Note that we cannot observe both of the potential outcomes.

Denoting treatment by Di, we will observe Yi(0) for the ith person if Di = 0, and similarly for Yi(1). If we denote observed Y for the ith person by Yi, we have

$Yi = Di * Yi(1) + (1 − Di) * Yi(0)$.

Consider the following treatment assignment:

```
D <- c(rep(1,4), rep(0,4))
```

The observed Ys with this treatment assignment are (Table 10.4):

```
Yi <- D*Y_po1 + (1-D)*Y_po0
surg_D <- data.frame(D,Yi)
surble <- xtable(surg_D, caption = "Treatment assignment and
        observed outcomes")
print(surble, caption.placement = "top")
```

We estimate the difference of means between treated and control groups:

```
lm(Yi ~ D)
##
## Call:
## lm(formula = Yi ~ D)
##
## Coefficients:
## (Intercept)            D
##        7.00        -2.75
```

We see that with this particular treatment assignment D, the average treatment effect is −2.75.

We can randomize assignment with:

```
sample(D, replace = FALSE)
## [1] 0 0 0 1 0 1 1 1
#once again, Ass is Assignment
Ass <- sample(D, replace = FALSE)
Ass
## [1] 1 0 1 0 0 0 1 1
```

So what will happen if we randomize assignment? We look at the sampling distribution of the estimated effect by using a loop.

```
iter <- 70
mean_effect <- numeric(iter)
for(i in 1: iter) {
   Ass <- sample(D, replace = FALSE)
   Out <- Ass*Y_po1 + (1 - Ass)*Y_po0
   mod_r <- lm(Out ~ Ass)
   mean_effect[i] <- mod_r$coeff[2]
}
round(mean(mean_effect),2)
## [1] -1.91
```

The mean of the different estimated effects is close to the true average effect (-2). We plot the sampling distribution.

```
pdoc <- data.frame(mean_effect)
ggplot(pdoc, aes(y = mean_effect)) +
        geom_boxplot() +
   coord_flip()
```

The boxplot (Fig. 10.6) shows the sampling distribution of the estimated effects; the histogram (Fig. 10.7) shows that the distribution has gaps and several peaks.

```
ggplot(pdoc, aes(x = mean_effect)) +
        geom_histogram(fill = "grey50") +
   geom_vline(xintercept =
       quantile(mean_effect, probs =
              c(0.25,0.5,0.75)),
      linetype = "dashed")
```

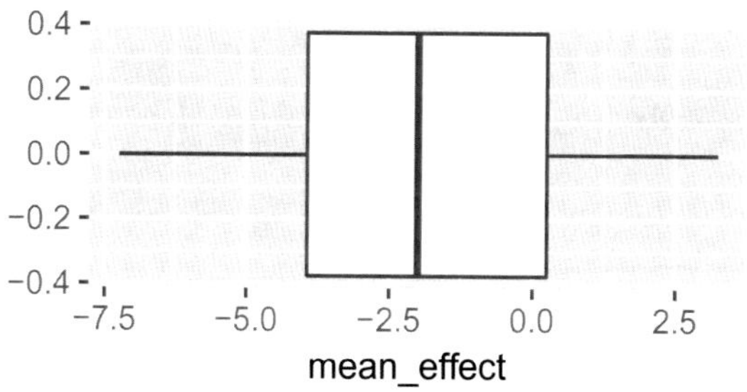

Fig. 10.6 Boxplot of estimated effect of surgery

Fig. 10.7 Histogram of
distribution of estimated
effects of surgery, lines show
first, second and third
quartiles

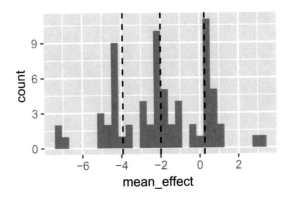

☞ **Your Turn**   A perfect doctor, or a doctor with perfect knowledge of
the potential outcomes, would assign surgery only to the
patients who will benefit from it. (Of course, if the doctor
had such perfect knowledge, we would not need to study
the effects of the surgery on people.) What will the average effect of the surgery be if the perfect doctor assigns
surgery?

10.2.4 Covariate Adjustment

Which covariates to adjust? Causal graphs are called directed acyclic graphs (DAGS).
According to Elwert (2013, p. 246), 'DAGS are visual representations of qualitative causal assumptions. ... DAGS are rigorous tools with formal rules for deriving
mathematical proofs. And yet, in many situations, using DAGs in practice requires
only modest formal training and some elementary probability training. DAGs are
thus extremely effective for presenting hard-won lessons of modern methodological
research in a language comprehensible to applied researchers.' Causal graphs are
particularly useful for illuminating issues of covariate adjustment.

If x causes y and a third variable is present, should we adjust for the third variable
in a regression of y on x? We use simulation to answer the question, and consider
three cases:

1. $y1 \leftarrow common\ cause \rightarrow x1$. Here the variable common cause causes both $y1$
   and $x1$.
2. $y2 \leftarrow intermediary \leftarrow x2$. Here $x2$ causes $y2$ through or via the intermediary
   variable.
3. $y3 \rightarrow collider \leftarrow x3$. Here $x3$ and $y3$ both cause the collider.

Some examples of possible common cause, intermediary and collider variables are:

1. Common cause. *More Murders* ← **Time** → *More antibiotics.*
2. Intermediary. *Heart disease* ← **Cholesterol** ← *Eating junk or overprocessed food.*
3. Collider. *Battery dead* → **car not starting** ← *No petrol.*

We could believe in these mechanisms, they may or may not be true.

We now turn to simulation, beginning with the common cause scenario: $y1 \leftarrow com.cause \rightarrow x1$.

```
com.cause <- runif(100, min = 10, max = 20)
x1 <- 2 * com.cause + rnorm(100,0,0.5)
y1 <- 2 * com.cause +  rnorm(100,0,0.5)
```

We regress (1) y1 on x1 and (2) y1 on x1 and the common cause.

```
library(texreg)
m1 <- lm(y1 ~ x1)
m2 <- lm(y1 ~ x1 + com.cause)
texreg(list(m1, m2), caption = "Common cause",
       caption.above = TRUE)
```

If we do not include the variable com.cause in the regression of y1 on x1 we get a spurious effect, which is statistically significant (Table 10.5).

We generate data for the intermediate variable scenario: $y2 \leftarrow inter \leftarrow x2$

```
x2 <- runif(100,min=10,max=20)
inter <- 2 * x2 + rnorm(100,0,0.5)
y2 <- 2 * inter +  rnorm(100,0,0.5)
inter1 <- lm(y2 ~ x2)
inter2 <- lm(y2 ~ x2 + inter)
```

Table 10.5 Common cause (True effect of x1 on y1 is 1)

	Model 1	Model 2
(Intercept)	−0.26	−0.39
	(0.35)	(0.26)
x1	1.01***	0.05
	(0.01)	(0.10)
com.cause		1.92***
		(0.21)
R^2	0.99	0.99
Adj. R^2	0.99	0.99
Num. obs.	100	100
RMSE	0.65	0.48

$^{*}p < 0.05$; $^{**}p < 0.01$; $^{***}p < 0.001$

Table 10.6 Intermediate variable (True effect of x2 on y2 is 4)

	Model 1	Model 2
(Intercept)	−0.64	0.03
	(0.52)	(0.27)
x2	4.06***	0.34
	(0.03)	(0.22)
inter		1.83***
		(0.11)
R^2	0.99	1.00
Adj. R^2	0.99	1.00
Num. obs.	100	100
RMSE	1.00	0.50

$^*p < 0.05; ^{**}p < 0.01; ^{***}p < 0.001$

```
texreg(list(inter1, inter2), caption = "Intermediate variables",
       caption.above = TRUE)
```

In this scenario, we should not control for the intermediate variable (Table 10.6).

☞ **Your Turn**   After generating y2, x2 and inter with the code above, regress y2 on x2, and y2 on x2 and inter. What do you observe?

We now turn to the collider scenario:
$y \rightarrow collider \leftarrow x$

```
x3 <- rnorm(100)
y3 <- rnorm(100)
collider <- 4 * y3 + 4 * x3 + 0.3 * rnorm(100)
```

$y \rightarrow collider \leftarrow x$

```
m5 <- lm(y3 ~ x3)
m6 <- lm(y3 ~ x3 + collider)
texreg(list(m5, m6), caption = "Collider",
       caption.above = TRUE)
```

In the collider case, we should not control for the collider (Table 10.7). We see that 'controlling' is not always good.

Good control   Controlling for the common cause is good.
Bad control   Controlling for the intermediary and collider variables is bad.

Table 10.7 Collider (True effect of x3 on y3 is 0)

	Model 1	Model 2
(Intercept)	0.01	−0.00
	(0.10)	(0.01)
x3	0.10	−1.00***
	(0.11)	(0.01)
collider		0.25***
		(0.00)
R^2	0.01	0.99
Adj. R^2	−0.00	0.99
Num. obs.	100	100
RMSE	0.99	0.07

$^{*}p < 0.05; ^{**}p < 0.01; ^{***}p < 0.001$

Authors like Gelman and Hill (2007) stress that we should not control for post-treatment variables. Here, we have shown that we should only control for variables that affect x, and not those affected by x. In the case of an intermediate variable, we can get the effect of y on inter, and x on inter, and from them get the effect of y on x. In certain situations this may be a useful strategy.

10.2.5 Selecting Regressors by Statistical Significance

This section seeks to illuminate the following argument by Freedman (1983, p. 152):

> When regression equations are used in empirical work, the ratio of data points to parameters is often low; furthermore, variables with small coefficients are often dropped and the equations refitted without them. ... Such practices can distort the significance levels of conventional statistical tests. The existence of this effect is well known, but its magnitude may come as a surprise, even to a hardened statistician.

We will generate a random variable y and 10 regressors, none of which are causally related to y.

```
set.seed(80)
x1 <- rnorm(30)
x2 <- rnorm(30)
x3 <- rnorm(30)
x4 <- rnorm(30)
x5 <- rnorm(30)
x6 <- rnorm(30)
x7 <- rnorm(30)
x8 <- rnorm(30)
x9 <- rnorm(30)
x10 <- rnorm(30)
y <-  rnorm(30)
mod1 <- lm(y ~ x1 + x2 + x3 + x4 + x5 + x6 + x7 + x8 + x9 + x10)
```

Table 10.8 Selection by statistical significance

	Model 1	Model 2
(Intercept)	0.08	0.02
	(0.17)	(0.14)
x1	0.09	
	(0.21)	
x2	0.28	0.22
	(0.19)	(0.15)
x3	0.06	
	(0.15)	
x4	0.09	
	(0.17)	
x5	0.02	
	(0.15)	
x6	0.19	
	(0.18)	
x7	0.19	
	(0.21)	
x8	0.13	
	(0.22)	
x9	−0.10	
	(0.22)	
x10	−0.36*	−0.37**
	(0.13)	(0.11)
R^2	0.44	0.33
Adj. R^2	0.15	0.28
Num. obs.	30	30
RMSE	0.83	0.76

$^*p < 0.05$; $^{**}p < 0.01$; $^{***}p < 0.001$

On the basis of Model 1 in Table 10.8, we select the regressors x2 and x10:

```
mod2 <- lm(y ~ x2 + x10)
texreg(list(mod1, mod2), caption = "Selection by statistical
    significance", caption.above = TRUE)
```

Although y is not related to any x variable, in Model 2 in Table 10.8, we get a high statistical significance for the coefficient of x10.

☞ **Your Turn**   Generate your data as above, changing the value inside set.seed() to 999. Run a regression of y on the 10 x variables. Select the three most statistically significant xs and run a second regression. What do you observe?

10.3 Experiments

10.3.1 Example: Anchoring

In his book, *Thinking, Fast and Slow*, Kahneman (2011, p. 119–120) explains what the anchoring effect is: 'If you are asked whether Gandhi was more than 114 years old when he died you will end up with a much higher estimate of his age at death than you would if the anchoring question referred to death at 35'.

We carried out an experiment that Kahneman and Tversky originally conducted with students at the University of Oregon (Kahneman 2011) to study anchoring. We asked the students to fill in the questions below.

- We chose (by computer) a random number between 0 and 100.
- The number selected and assigned to you is $X = ...$
- Do you think the percentage of countries, among all those in the United Nations, that are in Africa is higher or lower than X?............
- Give your best estimate of the percentage of countries, among all those in the United Nations, that are in Africa. ....

The number X was randomly assigned within a class, with a high X (65) being assigned to the treatment group, and a low X (10) being assigned to a control group. X was recorded as the variable `Prompt`.

The data are in the file `anchor`, containing the following variables:

- `Prompt`. High if $X = 65$, low if $X = 10$.
- `African`. Best guess of the percentage of countries among all those in the United Nations, that are in Africa.
- `Class`. Which of the two classes the students belonged to–IES or TERI.

We read in the data.

```
library(tidyverse)
anchor <- read_csv("anchor")
```

We create means for treatment and control groups, by class.

```
anchor %>%
  group_by(Class, Prompt) %>%
  summarize(count = n(),
            mean_AF = mean(African))
## # A tibble: 4 x 4
## # Groups:   Class [2]
##    Class Prompt  count mean_AF
##    <chr> <chr>   <int>   <dbl>
## 1 IES   High_65     4    58
## 2 IES   Low_10      4    14.8
## 3 TERI  High_65    16    24.5
## 4 TERI  Low_10     17    21.9
```

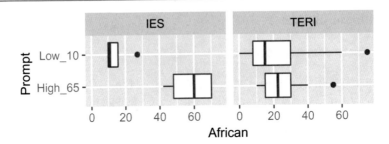

Fig. 10.8 Boxplots of Estimate of African countries in UN by high or low Prompt and by Class

In the IES class there was a large difference in means between the treatment and control groups (58 compared to 15), almost echoing the corresponding values of Prompt. In the TERI class there was a very small difference in means (also see Fig. 10.8).

```
ggplot(anchor, aes(y=African,
                   x = Prompt)) +
  geom_boxplot()  +
  facet_wrap(~ Class) +
  coord_flip()
```

We analyse the effects by Class. The IES class is small, only 8. For such a small class, we feel more comfortable using a permutation test (see Chap. 9) or randomization inference, which does not make distributional assumptions. Also, a small class helps us illustrate randomization inference in the case of experiments. We further zoom in on the first two units.

```
anchor_12 <- anchor %>%
  filter(Class == "IES") %>%
  dplyr::select(Prompt, African)
anchor_12[1:2,]
## # A tibble: 2 x 2
##    Prompt   African
##    <chr>      <dbl>
## 1 Low_10        27
## 2 High_65       70
```

The first person got a low prompt, for that person, the X was 10. The person estimated that the percentage of countries that were African was 27. The next person got a prompt (X) of 70; and estimated African as 70.

If we consider just these two people, we would get a treatment effect of 70−27, i.e. 43.

Another possible treatment assignment would be if the first person had got a high prompt and the second person had got a low prompt. But we don't observe the outcomes in this case. However under a sharp null hypothesis, the treatment effect for each person is zero and so the first person will guess African is 27 even if the person gets a low Prompt of 10. The second person would guess African is 70 even if receiving treatment of high prompt. Therefore the treatment effect will be

−43. We examine a large number of random assignments and calculate the average treatment effect under each assignment.

We will conduct randomization inference with the data for class IES.

```
anchor1 <- anchor %>%
  filter(Class == "IES")
```

The ri2 package conducts (Coppock 2019) randomization inference. We extract the outcome vector as Y and the treatment as Z.

```
library(ri2)
Y <- anchor1$African
Z <- ifelse(anchor1$Prompt == "High_65", 1, 0)
Z
## [1] 0 1 1 1 0 1 0 0
anchor_IES_P <- data.frame(Y, Z)
```

Having created a data frame containing Y and Z, we declare the random assignment: 4 of the 8 units were assigned to treatment.

```
declaration_1 <- declare_ra(N = 8, m = 4)

ri2_IES_P <- conduct_ri(Y ~ Z,
  data = anchor_IES_P,
  declaration = declaration_1
)
```

We then conduct randomization inference.

```
summary(ri2_IES_P)
##    term estimate two_tailed_p_value
## 1     Z    43.25          0.02857143
```

The estimate of the effect is 43, with a p-value of 0.03 (Fig. 10.9).

```
plot(ri2_IES_P)
```

We now analyse the data for the larger class, TERI.

```
anchor2 <- anchor %>%
  filter(Class == "TERI")

Y <- anchor2$African
Z <- ifelse(anchor2$Prompt == "High_65", 1, 0)

anchor_TERI <- data.frame(Y, Z)


declaration2 <- declare_ra(N = 33, m = 16)

ri2_TERI <- conduct_ri(Y ~ Z,
                 data = anchor_TERI,
                 declaration = declaration2)
```

Fig. 10.9 Randomization inference distribution and estimate of anchoring effect in class IES

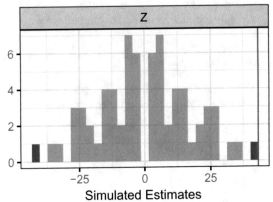

Fig. 10.10 Randomization inference distribution and estimate of anchoring effect in class TERI

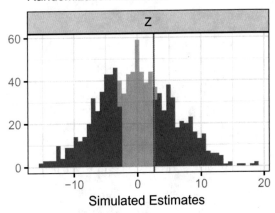

```
summary(ri2_TERI)
##   term estimate two_tailed_p_value
## 1    Z 2.605882              0.675
```

The estimate for TERI is 2.6 with a p-value of 0.67 (Fig. 10.10).

```
plot(ri2_TERI)
```

We get similar results using a linear model (Table 10.9).

Table 10.9 Dependent variable is African, class TERI

	Model 1
(Intercept)	24.50***
	(4.28)
PromptLow_10	−2.61
	(5.97)
R^2	0.01
Adj. R^2	−0.03
Num. obs.	33
RMSE	17.14

$^*p < 0.05;\ ^{**}p < 0.01;\ ^{***}p < 0.001$

```
mod2 <- lm(African ~ Prompt, data = anchor2)
library(texreg)
texreg(list(mod2), caption = "Dependent variable is African,
       class TERI", caption.above = TRUE)
```

☞ **Your Turn** Print out the Kahneman anchoring questions, and make a few copies. Ask a small group of people you know to fill out the questions. Record the data, and then analyse.

10.3.2 Example: Women as Policymakers

Duflo and Chattopadhyay (2004) studied the effect of reserving positions of leadership in Village Councils in India on the kinds of projects undertaken by them. A subset of their data is presented in Imai (2018), this corresponds to data for Birbhum district in the state of West Bengal. The data can be accessed as follows (remove the hash symbol):

```
#library("devtools")
#install_github("kosukeimai/qss-package", build_vignettes = TRUE)
#data(women, package = "qss")
```

We have already got the data in our computer, so we read it in.

```
library(tidyverse)
women <- read.csv("~/Documents/R/ies2018/women.csv")
```

The variables in the dataset women of interest to us are:

- GP identifier for the Gram Panchayat.
- village identifier for each village.
- female whether the GP had a female leader or not.

Fig. 10.11 Boxplots of
water in treated and control
groups

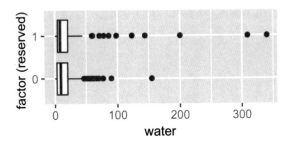

- `water` number of new or repaired drinking water facilities in the village since
  the reservation policy started.

The reservation is done by random assignment.

```
women %>%
  group_by(reserved) %>%
  summarize(count_res = n(),
            mean_female = mean(female),
            mean_water = mean(water))
## # A tibble: 2 x 4
##    reserved count_res mean_female mean_water
##       <int>     <int>       <dbl>      <dbl>
## 1         0       214      0.0748       14.7
## 2         1       108      1            24.0
```

The mean for water in the treatment group is 24, in the control group it is 15.

```
ggplot(women,aes(y = water, x = factor(reserved))) +
  geom_boxplot() +
  coord_flip()
```

Figure 10.11 shows that some of the treated villages had very high levels of water,
a feature of the sample we observe thanks to the boxplots.

The data is clustered, so we incorporate this while carrying out randomization
inference with the ri2 package.

```
dat <- data.frame(Y = women$water,
                  Z = women$reserved ,
                  cluster = women$GP)
head(dat)
##    Y Z cluster
## 1 10 1       1
## 2  0 1       1
## 3  2 1       2
## 4 31 1       2
## 5  0 0       3
## 6  0 0       3
declaration <- with(dat, {
  declare_ra(clusters = cluster
             )
```

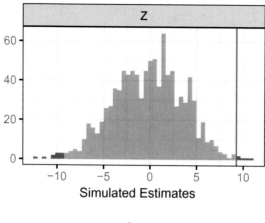

Fig. 10.12 Randomization inference distribution for reserved

```
})

declaration
## Random assignment procedure: Cluster random assignment
## Number of units: 322
## Number of clusters: 161
## Number of treatment arms: 2
## The possible treatment categories are 0 and 1.
## The number of possible random assignments is approximately
    infinite.
## The probabilities of assignment are constant across units:
## prob_0 prob_1
##    0.5    0.5
ri2_out <- conduct_ri(
  Y ~ Z,
  sharp_hypothesis = 0,
  declaration = declaration,
  data = dat
)

summary(ri2_out)
##   term estimate two_tailed_p_value
## 1    Z 9.252423              0.015


plot(ri2_out)
```

We get an estimate of 9.25 with a low p-value of 0.015 (Fig. 10.12).

Table 10.10 Effect of women on water

	Model 1
(Intercept)	14.74
	[11.70; 17.77]
factor(reserved) 1	9.25
	[−0.79; 19.29]
R^2	0.02
Adj. R^2	0.01
Num. obs.	322
RMSE	33.45

We also use the `estimatr` package, and a linear model, and estimate cluster robust standard errors. Table 10.10 shows the resulting coefficients and confidence intervals.

```
women$reserved <- factor(women$reserved)
library(estimatr)
mod_water_r <- lm_robust(water ~ factor(reserved),
     clusters = GP, data = women)
library(texreg)
texreg(list(mod_water_r),
     caption = "Effect of women on water",
     ci.force = T, ci.test = NULL,
     caption.above = TRUE)
```

10.3.3 Example: Educational Programme

We look at the data from an experiment conducted around 1970, presented in Gelman and Hill (2007, p. 174–181). The outcome was reading test scores, and the treatment was exposure to an education television programme.

```
electric <- read.table("electric.dat", header = T) #str(electric)
```

The key variables in the dataset `electric` are:

- `Grade`. Grade of the student.
- `treated.Pretest`. Pre-test scores of the treated students.
- `control.Pretest`. Pre-test scores of the control students.
- `treated.Posttest`. Post-test scores of the treated students.
- `control.Posttest`. Post-test scores of the control students.

We have to wrangle the data, we need to combine the Pretest variables for treated and control students, similarly for Posttest, and create an indicator for treatment.

☞ **Your Turn**   Find the electric data on the website for the Gelman and Hill book. Work through the code below to wrangle the data.

```
post.test <-  c(electric$treated.Posttest,
    electric$control.Posttest)
pre.test <-
  c(electric$treated.Pretest,
    electric$control.Pretest)
grade <- rep(electric$Grade,2)
grade <- factor(grade)
rep(c(1,0),rep(3,2))
## [1] 1 1 1 0 0 0
treatment <- rep(c(1,0),
    rep(length(electric$treated.Posttest),2))
treatment <- factor(treatment)
n <- length(post.test)
elec <- tibble(post.test,
    pre.test,grade,treatment)
elec
## # A tibble: 192 x 4
##    post.test pre.test grade treatment
##        <dbl>    <dbl> <fct> <fct>
## 1      48.9     13.8  1      1
## 2      70.5     16.5  1      1
## 3      89.7     18.5  1      1
## 4      44.2      8.8  1      1
## 5      77.5     15.3  1      1
## 6      84.7     15    1      1
## 7      78.9     19.4  1      1
## 8      86.8     15    1      1
## 9      60.8     11.8  1      1
## 10     75.7     16.4  1      1
## # ... with 182 more rows
```

We will focus on grade1, we filter the data.

```
library(tidyverse)
elec_1 <- elec %>%
  filter(grade==1)
```

We plot boxplots of post-test scores of treated versus control group (Fig. 10.13).

```
ggplot(elec_1, aes(y = post.test,
  x = treatment)) +
    geom_boxplot() +
  coord_flip()
```

A scatter plot of post-test versus pre-test works well in this case (Fig. 10.14).

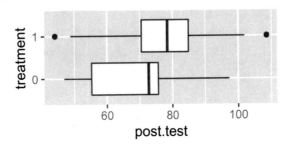

Fig. 10.13 Boxplots of post-test scores of treated versus control group

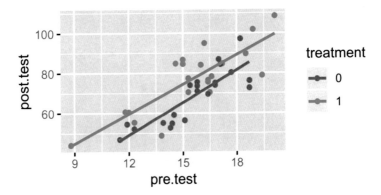

Fig. 10.14 Scatter plot of post-test versus pre-test scores for treatment and control groups

```
ggplot(elec_1, aes(x = pre.test,
  y = post.test,
  colour = treatment)) +
  geom_point() +
  stat_smooth(method=lm,
              se = FALSE)
```

Because this is an experiment, we need not control for pre-test scores, but including pre-test scores gives more precise estimates.

```
mod1.1 <- lm(post.test ~ treatment,
             data = elec_1)
mod1.2 <- lm(post.test ~ pre.test +
             treatment, data= elec_1)
texreg(list(mod1.1, mod1.2),
       ci.force = TRUE, ci.test = NULL,
       caption = "Effect of programme on scores",
       caption.above = TRUE)
```

In Table 10.11, the estimate of the coefficient of treatment is very similar in the two models. However, the estimate is more precise (narrower confidence interval) in Model 2.

Table 10.11 Effect of programme on scores

	Model 1	Model 2
(Intercept)	68.79	−11.02
	[62.38; 75.20]	[−28.24; 6.20]
treatment1	8.30	8.79
	[−0.76; 17.36]	[3.67; 13.91]
pre.test		5.11
		[4.03; 6.19]
R^2	0.07	0.71
Adj. R^2	0.05	0.70
Num. obs.	42	42
RMSE	14.98	8.46

☞ **Your Turn** Analyse the data for grades 2 and 4. How do the results compare with those for grade 1?

10.3.4 Example: Star

The Star project was a large experiment conducted in Tennessee in the United States. Three treatments were assigned at the classroom level: small classes (13–17 students), regular classes (22–25 students), and regular classes with an aide who would work with the teacher. The analysis here follows the presentation in the econometrics textbook by Hill et al. (2018), and focuses on the small class treatment in comparison with the control group of regular sized classes.

```
library(tidyverse)
library(POE5Rdata)
data(star)
#str(star)
```

The key variables in the dataset `star` are:

- `totalscore`. Reading plus math score.
- `small`. Is 1 if student was assigned to a small class.
- `boy, white-asian, freelunch`. Student descriptors.
- `tchexper`. Teacher's experience.

One of the treatments was using a teaching aide; we ignore these observations and focus on the small versus regular comparison.

Fig. 10.15 Boxplots of
totalscore for small and
regular classes

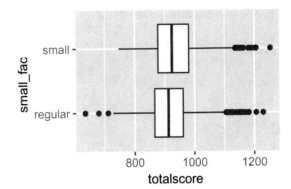

```
star <- star %>%
  filter(aide == 0) %>%
  dplyr::select(totalscore, small, tchexper,
        boy, freelunch, white_asian, schid) %>%
  mutate(small_fac = ifelse(small == 1, "small", "regular"),
        sch_fac = factor(schid))
#str(star)
star <- as_tibble(star)
```

```
star %>%
 group_by(small_fac)  %>%
  summarize(mscore = mean(totalscore),
            sdscore = sd(totalscore))
## # A tibble: 2 x 3
##    small_fac mscore sdscore
##    <chr>      <dbl>   <dbl>
## 1 regular     918.    73.1
## 2 small       932.    76.4
```

The mean score for the regular class was 918, for the small class it was 932.

```
ggplot(star, aes(x = small_fac, y = totalscore)) +
  geom_boxplot() +
  coord_flip()
```

In Fig. 10.15 we see that the distribution of totalscore shifts rightward for the group of students in small classes. We also see that in our sample, the lower outliers are not there in the small classes, which we notice thanks to the boxplots.

We can see how different covariates vary between the regular and small classes.

```
star %>%
 group_by(small_fac)  %>%
  summarize(mboy = mean(boy),
            mlunch = mean(freelunch),
            mw_a = mean(white_asian),
            mexper = mean(tchexper))
```

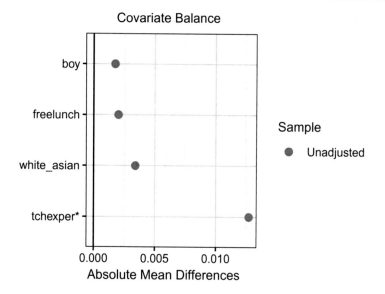

Fig. 10.16 Covariate balance for star data

```
## # A tibble: 2 x 5
##    small_fac  mboy mlunch  mw_a mexper
##    <chr>     <dbl>  <dbl> <dbl>  <dbl>
## 1 regular   0.513  0.474 0.681   9.07
## 2 small     0.515  0.472 0.685   9.00
```

The differences in means of covariates between treated and control groups are small. The `cobalt` package (Greifer 2019) helps us assess covariate balance (Fig. 10.16).

```
library(cobalt)
love.plot(small ~ boy + freelunch +
          white_asian + tchexper,
       data = star, stars = "std")
```

```
## Note: estimand and s.d.denom not specified; assuming
ATE and pooled.
```

We can conduct a formal test of balance as follows using a linear probability model (Hill et al. 2018).

```
mod_star_check <- lm(small ~ boy + white_asian + tchexper +
                freelunch, data = star)
library(texreg)
texreg(mod_star_check, ci.force = TRUE,
       ci.test = NULL, caption = "Checking balance",
       caption.above = TRUE)
```

Table 10.12, made with the `texreg` package (Leifeld 2013), confirms that treated and control groups are balanced.

Table 10.12 Checking balance

	Model 1
(Intercept)	0.47 [0.42; 0.52]
boy	0.00 [−0.03; 0.03]
white_asian	0.00 [−0.03; 0.04]
tchexper	−0.00 [−0.00; 0.00]
freelunch	−0.00 [−0.04; 0.03]
R^2	0.00
Adj. R^2	−0.00
Num. obs.	3743
RMSE	0.50

```
mod_star_1 <- lm(totalscore ~ small_fac, data = star)
mod_star_2 <- lm(totalscore ~ small_fac + boy +
    freelunch + white_asian, data = star)

mod_star_3 <- lm(totalscore ~ small_fac + boy +
    freelunch + white_asian + tchexper + sch_fac, data = star)

texreg(list(mod_star_1, mod_star_2, mod_star_3),
        omit.coef = "Intercept|sch_fac",
      ci.force= TRUE, ci.test = NULL,
      caption = "Effect of small class on total scores",
      caption.above = TRUE)
```

The effect of small classes (about 14 in model 1 and model 2 in Table 10.13) is stable in the three specifications. In model 3 we get a higher estimate and greater precision.

10.4 Matching

The basic idea of matching is simple. In an observational study, the treated and control units may differ greatly on some covariate. For example, units A and B, where A is treated and B is control, may differ on age and gender. However, C, a control unit, may be like A with respect to age and gender. We then match A to C, and use the comparison between matched units to assess the effect of treatment.

In practice, with several covariates, many of which may be continuous, we cannot match exactly, but have to use a more intricate method.

We first match the data, not looking at the outcome data, check for balance and then estimate the treatment effect.

Table 10.13 Effect of small class on total scores

	Model 1	Model 2	Model 3
small_facsmall	13.90	13.81	16.06
	[9.10; 18.69]	[9.23; 18.40]	[11.89; 20.22]
boy		−15.64	−13.46
		[−20.22; −11.06]	[−17.56; −9.35]
freelunch		−34.19	−36.34
		[−39.29; −29.09]	[−41.24; −31.43]
white_asian		12.58	25.26
		[7.11; 18.05]	[16.61; 33.91]
tchexper			0.82
			[0.40; 1.25]
R^2	0.01	0.09	0.30
Adj. R^2	0.01	0.09	0.29
Num. obs.	3743	3743	3743
RMSE	74.65	71.43	63.28

Matching details can be somewhat technical and so we rely on simulated data to get intuition about matching.

10.4.1 Simple Example with Synthetic Data

The basic idea of matching is quite intuitive. We look at a small simulated example. Assume that x, a binary variable, and w are causes of y. Also w affects y nonlinearly.

```
library(tidyverse)
```

We generate the data.

```
x <- c(rep(0,6),rep(1,6))
w <- c(30,18,20,10,10,17,20,18,10,10,17,3)
y <- (10 * x) + w + (0.2 * w^2) +
  (3 * (rnorm(12,1,1)))
wsq <- w^2
dat_mat <- data.frame(y,x,w,wsq)
dat_mat
##              y x  w wsq
## 1   218.38954 0 30 900
## 2    85.06152 0 18 324
## 3    99.19242 0 20 400
## 4    32.39997 0 10 100
## 5    37.37473 0 10 100
## 6    77.01007 0 17 289
## 7   109.19882 1 20 400
## 8    91.67818 1 18 324
## 9    37.42884 1 10 100
## 10   41.30772 1 10 100
```

Fig. 10.17 Scatter plot of y
versus w, unmatched data

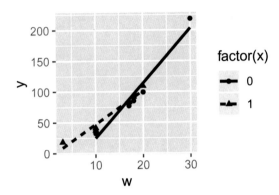

```
## 11   83.10518 1 17 289
## 12   18.43934 1  3   9
# tibble interferes with matching
```

Notice in `dat_mat` that the first observation has a value of w = 30, with x = 0 and the 12th has a value of w = 3. This lack of overlap can be seen in Fig. 10.17.

```
ggplot(dat_mat, aes(x = w, y = y,
            shape = factor(x),
            linetype = factor(x))) +
   geom_point() +
   geom_smooth(method = "lm", se = FALSE, col = "black")
```

We run three different regressions:

```
library(texreg)
mod1 <- lm(y ~ x + w + wsq, data = dat_mat)
mod2 <- lm(y ~ x + w, data = dat_mat)
mod3 <- lm(y ~ x , data = dat_mat)
texreg(list(mod1, mod2, mod3),
        caption = "Effect of omitting w and wsq",
          ci.force = T, ci.test = NULL,
        caption.above = TRUE)
```

If we regress y on x, w and wsq our estimate of the effect of y on x is close to the true estimate (Table 10.14). Even regressing y on x and w gives us a reasonable estimate. But the simple regression of y on x gives a very biased estimate.

We now use exact matching to match observations for which x = 0 with corresponding observations for which x = 1 with exactly the same ws. We use the `MatchIt` package, and `matchit()`, with method = "exact".

```
library(MatchIt)
match.1 <- matchit(x ~ w, data = dat_mat,
    method = "exact", replace = FALSE)
match.1
##
## Call:
```

Table 10.14 Effect of omitting w and wsq (True effect of x on y = 10)

	Model 1	Model 2	Model 3
(Intercept)	11.46	−38.48	91.57
	[5.94; 16.97]	[−68.28; −8.68]	[48.26; 134.89]
x	6.05	5.40	−28.05
	[3.51; 8.59]	[−15.01; 25.80]	[−89.30; 33.21]
w	−0.11	7.43	
	[−0.75; 0.54]	[5.92; 8.95]	
wsq	0.23		
	[0.21; 0.25]		
R^2	1.00	0.92	0.07
Adj. R^2	1.00	0.90	-0.02
Num. obs.	12	12	12
RMSE	2.11	17.00	54.13

Fig. 10.18 Covariate balance

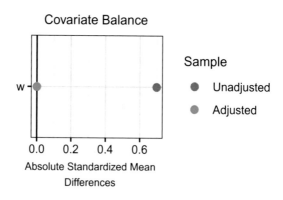

```
## matchit(formula = x ~ w, data = dat_mat, method = "exact",
##          replace = FALSE)
##
## Exact Subclasses: 4
##
## Sample sizes:
##            Control  Treated
## All              6        6
## Matched          5        5
## Unmatched        1        1
```

Exact matching has led to 5 of the treatment group observations being matched controls; 1 went unmatched.

The love.plot function in the cobalt package plots the covariate balance before and after matching (Fig. 10.18).

```
library(cobalt)
love.plot(match.1, stars = "std")
```

Table 10.15 Regressions with matched data (True effect of x on y = 10)

	Model 1	Model 2	Model 3
(Intercept)	2.87	−31.79	66.21
	[−30.62; 36.37]	[−38.30; −25.28]	[39.25; 93.16]
x	6.34	6.34	6.34
	[3.53; 9.14]	[2.95; 9.72]	[−31.78; 44.45]
w	1.30	6.53	
	[−3.70; 6.30]	[6.13; 6.94]	
wsq	0.18		
	[0.01; 0.35]		
R^2	1.00	0.99	0.01
Adj. R^2	0.99	0.99	−0.11
Num. obs.	10	10	10
RMSE	2.26	2.73	30.75

We now extract the matched data.

```
match_dat <- match.data(match.1)
match_dat
##              y x  w wsq weights subclass
## 2    85.06152 0 18 324       1        2
## 3    99.19242 0 20 400       1        1
## 4    32.39997 0 10 100       1        3
## 5    37.37473 0 10 100       1        3
## 6    77.01007 0 17 289       1        4
## 7   109.19882 1 20 400       1        1
## 8    91.67818 1 18 324       1        2
## 9    37.42884 1 10 100       1        3
## 10   41.30772 1 10 100       1        3
## 11   83.10518 1 17 289       1        4
```

We run the same regressions that we ran earlier. With exact matching, the estimate across the regression specifications is the same (Table 10.15).

```
mod_m1 <- lm(y ~ x + w + wsq, data = match_dat)
mod_m2 <- lm(y ~ x + w, data = match_dat)
mod_m3 <- lm(y ~ x , data = match_dat)
texreg(list(mod_m1, mod_m2, mod_m3),
        caption = "Regressions with matched data",
        ci.force = T, ci.test = NULL,
        caption.above = TRUE)

ggplot(match_dat, aes(x = w, y = y,
            shape = factor(x),
            linetype = factor(x))) +
    geom_point() +
    geom_smooth(method = "lm", se = FALSE, col = "black")

rm(list=ls())
```

Fig. 10.19 Scatter plot of y against w after matching

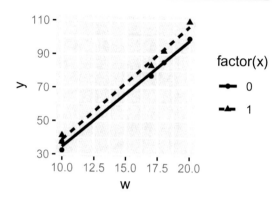

We can compare our earlier Fig. 10.17 with the unmatched data with Fig. 10.19 for the matched data. Ho et al. (2011) advocate matching as a non-parametric method to reduce model dependence (as in Table 10.15).

10.4.2 Example: Labour Training Programme

Lalonde (1986) showed that the experimental and econometric evaluation (with observational data) of a labour training programme—the National Supported Work Demonstration—reached different conclusions. However, Dehejia and Wahba (1999) subsequently used matching; they showed that an observational study could arrive at the results of the experimental study.

We will work with the version of the data available in the MatchIt package.

```
data(lalonde, package = "MatchIt")
#write.csv(ll,"ll.csv")
#ll <- read.csv("ll.csv")
str(lalonde)
## 'data.frame':    614 obs. of  10 variables:
##  $ treat   : int  1 1 1 1 1 1 1 1 1 1 ...
##  $ age     : int  37 22 30 27 33 22 23 32 22 33 ...
##  $ educ    : int  11 9 12 11 8 9 12 11 16 12 ...
##  $ black   : int  1 0 1 1 1 1 1 1 1 0 ...
##  $ hispan  : int  0 1 0 0 0 0 0 0 0 0 ...
##  $ married : int  1 0 0 0 0 0 0 0 0 1 ...
##  $ nodegree: int  1 1 0 1 1 1 0 1 0 0 ...
##  $ re74    : num  0 0 0 0 0 0 0 0 0 0 ...
##  $ re75    : num  0 0 0 0 0 0 0 0 0 0 ...
##  $ re78    : num  9930 3596 24909 7506 290 ...
```

The outcome variable is the earnings in 1978, re78. The treatment is treat. There are demographic covariates, and past earnings in 1974 re74 and 1975 re75.

We rename lalonde to save typing.

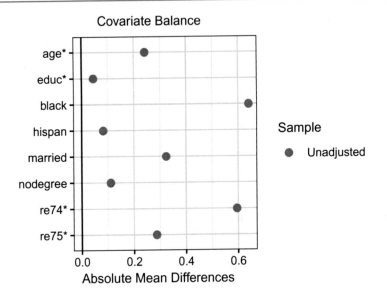

Fig. 10.20 Covariate balance in l1 before matching

```
l1 <- lalonde
```

```
love.plot(treat ~ age + educ + black + hispan + married +
    nodegree + re74 + re75, data = l1, stars="std")
```

The covariates are imbalanced, especially black and re74 (Fig. 10.20).

```
mod_la1 <- lm(re78 ~ treat, data = l1)
mod_la2 <- lm(re78 ~ treat + age + educ + black + hispan + married +
    nodegree + re74 + re75, data = l1)
texreg(list(mod_la1, mod_la2),
        caption = "Regression with lalonde data",
        ci.force = T, ci.test = NULL)
```

In Table 10.16, Model 1 and Model 2 give us very different estimates.

We now use genetic matching, which incorporates an algorithm to improve balance (Sekhon 2011). According to Sekhon (2011, p. 3), 'A significant shortcoming of common matching methods such as Mahalanobis distance and propensity score methods is that they may (and in practice, frequently do) make balance worse across measured potential confounders'. The advantage of genetic matching is that it (Sekhon 2011, p. 1) 'automatically finds the set of matches which minimize the discrepancy between the distribution of potential confounders in the treated and control groups—i.e. the covariate balance is maximized'. The MatchIt package uses the Matching package for genetic matching.

Table 10.16 Regression with lalonde data

	Model 1	Model 2
(Intercept)	6984.17 [6277.19; 7691.15]	66.51 [−4709.42; 4842.45]
treat	−635.03 [−1922.99; 652.94]	1548.24 [16.96; 3079.52]
age		12.98 [−50.70; 76.65]
educ		403.94 [92.49; 715.39]
black		−1240.64 [−2747.39; 266.11]
hispan		498.90 [−1347.28; 2345.07]
married		406.62 [−956.48; 1769.72]
nodegree		259.82 [−1401.14; 1920.77]
re74		0.30 [0.18; 0.41]
re75		0.23 [0.03; 0.44]
R^2	0.00	0.15
Adj. R^2	−0.00	0.14
Num. obs.	614	614
RMSE	7471.13	6947.92

```
set.seed(123)
match.l1 <- matchit(treat ~ age + educ + black + hispan + married +
    nodegree + re74 + re75,
        data = l1, method = "genetic",
      replace = FALSE, pop.size = 50, print = 0)#, caliper = 0.4)
```

Loading required namespace: rgenoud

```
    #print = 0)
match.l1
##
## Call:
## matchit(formula = treat ~ age + educ + black + hispan + married +
##     nodegree + re74 + re75, data = l1, method = "genetic",
##     replace = FALSE, pop.size = 50, print = 0)
##
## Sample sizes:
##           Control Treated
## All           429     185
## Matched       185     185
```

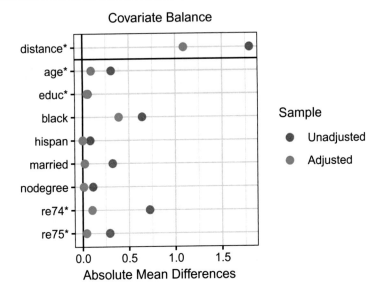

Fig. 10.21 Covariate balance in l1 before and after matching

```
## Unmatched      244         0
## Discarded        0         0
```

All 185 treated observations were matched.

```
love.plot(match.l1, stars ="std")
```

Covariate balance has improved after matching (Fig. 10.21).

```
match_dat <- match.data(match.l1)
```

```
ggplot(l1, aes(x = factor(treat),
  y = re78)) +
  geom_boxplot() + coord_flip()
```

```
ggplot(match_dat, aes(x = factor(treat),
  y = re78)) +
  geom_boxplot() + coord_flip()
```

In the unmatched data the median and the 75th percentile values of re78 are higher in the control group, the opposite is true in matched data (Figs. 10.22 and 10.23). Also, note the outliers in the treated and control groups.

```
mod_la_match1 <- lm(re78 ~ treat, data = match_dat)
mod_la_match2 <- lm(re78 ~ treat + age + educ + black + hispan +
    married + nodegree + re74 + re75, data = match_dat)
texreg(list(mod_la_match1, mod_la_match2),
```

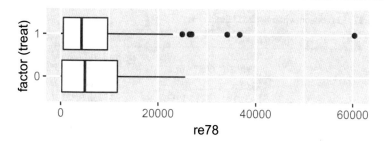

Fig. 10.22 Boxplots of treated and control observations in unmatched data

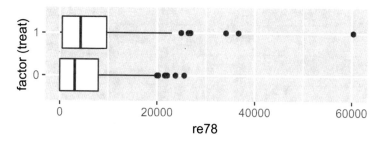

Fig. 10.23 Boxplots in matched data

```
caption = "Regression with matched lalonde data",
   ci.force = T, ci.test = NULL)
```

In Table 10.13 we see that with the matched data we get a treatment effect of 756 by estimating the difference of means. We get an estimate of 1220 once we include controls in the regression (Table 10.17).

10.4.3 Sensitivity Analysis

In an experiment, treatment is randomly assigned. As a result of the random assignment, treatment and control units are balanced with respect to covariates observed and unobserved.

In an observational study, on the other hand, treatment is not randomly assigned. We match to bring about balance between treatment and control units with respect to observed covariates; but unobserved covariates may be influencing treatment.

Although we cannot observe what is unobserved, we can conduct a sensitivity analysis. The following paragraph by Rosenbaum (2005, p. 1809) summarizes sensitivity analysis:

> The sensitivity analysis imagines that in the population before matching or stratification, subjects are assigned to treatment or control independently with unkown probabilities. Specifically, two subjects who look the same at baseline before treatment—that is, two subjects with the same observed covariates—may nonetheless differ in terms of unobserved covariates, so

Table 10.17 Regression with matched lalonde data

	Model 1	Model 2
(Intercept)	5593.26	−1150.87
	[4549.36; 6637.16]	[−7660.86; 5359.13]
treat	755.88	1220.33
	[−720.41; 2232.18]	[−375.69; 2816.35]
age		7.89
		[−78.16; 93.94]
educ		597.52
		[169.16; 1025.87]
black		−1139.82
		[−2923.53; 643.89]
hispan		278.15
		[−3013.04; 3569.34]
married		940.61
		[−1042.30; 2923.51]
nodegree		79.29
		[−2139.55; 2298.12]
re74		0.03
		[−0.17; 0.23]
re75		0.35
		[0.03; 0.67]
R^2	0.00	0.09
Adj. R^2	0.00	0.07
Num. obs.	370	370
RMSE	7244.29	6994.93

that one subject has an odds of treatment that is up to $\Gamma \geq 1$ greater than the odds for another subject. In the simplest randomized experiment, everyone has the same chance of receiving the treatment, so $\Gamma = 1$. If $\Gamma = 2$ in an observational study, one subject might be twice as likely as another to receive the treatment because of unobserved pre-treatment differences. The sensitivity analysis asks how much hidden bias can be present—that is, how large can Γ be—before the qualitative conclusions of the study begin to change. A study is highly sensitive to hidden bias if the conclusions change for Γ just barely larger than 1, and it is insensitive if the conclusions change only for quite large values of Γ.

The R package `rbounds` (Keele 2014) draws on the package `Matching` to do genetic matching and then conducts a sensitivity analysis. We will do this for the lalonde data in the MatchIt package that we had analysed earlier.

We need to provide the data to `Matching` in a certain format. Y below is for the outcome, Tr for treatment, X for covariates.

```
library(rbounds)


Y <- l1$re78
Tr <- l1$treat #the treatment of interest
```

```
X <- cbind(11$age, 11$educ, 11$black, 11$hispan,
   11$married, 11$nodegree, 11$re74, 11$re75)

BalanceMat <- cbind(11$age, I(11$age^2), 11$educ, I(11$educ^2),
11$black, 11$hispan, 11$married,
11$nodegree, 11$re74 , I(11$re74^2), 11$re75, I(11$re75^2),
  I(11$re74*11$re75),
I(11$age*11$nodegree), I(11$educ*11$re74), I(11$educ*75))
```

The `GenMatch()` function carries out genetic matching.

```
#Genetic Weights
gen1 <- GenMatch(Tr=Tr, X=X, BalanceMat=BalanceMat, pop.size=50,
data.type.int=FALSE, print=0, replace=FALSE)

#Match
mgen1 <- Match(Y=Y, Tr=Tr, X=X, Weight.matrix=gen1, replace=FALSE)
summary(mgen1)
##
## Estimate...   765.56
## SE........   699.76
## T-stat.....   1.094
## p.val......   0.27394
##
## Original number of observations..............   614
## Original number of treated obs..............   185
## Matched number of observations..............   185
## Matched number of observations  (unweighted).   185
```

The difference of means estimate provided by Matching is 765, with an Abadie-Imbens standard error of 699. The `psens()` function provides a sensitivity analysis: if Gamma (Γ) changes, how does the Wilcoxon Signed Rank p-value change? Even with Gamma $= 1$, the p-value is well above 0.05 or 0.1. There is a large sensitivity to possible hidden bias due to missing covariates.

```
psens(mgen1, Gamma = 1.5, GammaInc = 0.1)
##
##   Rosenbaum Sensitivity Test for Wilcoxon Signed Rank P-Value
##
## Unconfounded estimate ....   0.274
##
##   Gamma Lower bound Upper bound
##    1.0       0.2740       0.2740
##    1.1       0.1280       0.4732
##    1.2       0.0519       0.6627
##    1.3       0.0187       0.8074
##    1.4       0.0061       0.9006
##    1.5       0.0019       0.9530
##
##   Note: Gamma is Odds of Differential Assignment To
##   Treatment Due to Unobserved Factors
##
```

The `hlsens()` provides a sensitivity analysis for the Hodges–Lehmann point estimate. With a low Gamma of 1.1, the Lower Bound and the Upper Bound are −0.017 and 547, respectively.

```
hlsens(mgen1, Gamma = 1.5, GammaInc = 0.1)
##
##  Rosenbaum Sensitivity Test for Hodges-Lehmann Point Estimate
##
## Unconfounded estimate .... 401.3827
##
##  Gamma Lower bound Upper bound
##    1.0  401.380000     401.38
##    1.1   -0.017264     547.18
##    1.2 -211.320000     910.18
##    1.3 -443.520000    1160.60
##    1.4 -702.620000    1382.90
##    1.5 -959.620000    1581.90
##
##  Note: Gamma is Odds of Differential Assignment To
##  Treatment Due to Unobserved Factors
##
```

10.4.4 Example: Lead Exposure

Rosenbaum (2017, p. 216) writes:

> Matching may use technical tools to balance many observed covariates, x, but it leaves in its wake a simple structure, perhaps matched pairs, in which treated and control groups are readily seen to be comparable in terms of each measured covariate. With concerns about the measured covariates removed from the picture, our attention turns to the challenging issues that determine whether or not an observational study is convincing.

Does a father's exposure to lead at work affect his children? What kinds of comparisons will shed light on this? In the example presented in Rosenbaum (2017), the fathers worked in a battery manufacturing plant in Oklahoma.

The data is in the DOS package, and the observations are already matched.

```
data(lead, package = "DOS")
head(lead)
##   control exposed level  hyg    both dif
## 1      13      14  high good high.ok   1
## 2      16      13  high good high.ok  -3
## 3      11      25  high good high.ok  14
## 4      18      41  high  mod high.ok  23
## 5      24      18  high  mod high.ok  -6
## 6       7      49  high  mod high.ok  42
```

Our first comparison is of children whose fathers worked in the battery plant with matched control children.

Fig. 10.24 Boxplots of lead
levels of children whose
fathers worked in the battery
plant (exposed) compared to
matched control children of
same age

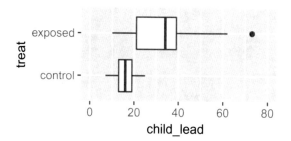

```
child_lead <- c(lead$control, lead$exposed)
treat <- c(rep("control",33), rep("exposed",33))
child_lead_dat <- data.frame(child_lead, treat)

ggplot(child_lead_dat,
       aes(x = treat, y = child_lead)) +
  geom_boxplot() +
  ylim(0,80) +
  coord_flip()
```

Figure 10.24 shows that the distribution of lead levels is far greater for the exposed
children.

Our second comparison is of children whose fathers had different levels of exposure to lead at the battery plant.

```
library(forcats)
llevel <- c(
  "low", "medium", "high"
)

lead$F_level <- factor(lead$level,
     levels = llevel)


ggplot(lead,
       aes(x = F_level, y = exposed)) +
  geom_boxplot() +
  ylim(0,80) + coord_flip()
```

Children whose fathers had higher levels of exposure had higher levels of lead
(Fig. 10.25).

Our third comparison is between children in the control group separated on the
basis of the matched treated child's father's exposure.

```
ggplot(lead,
       aes(x = F_level, y = control)) +
  geom_boxplot() +
  ylim(0,80) + coord_flip()
```

We can see that the effect evident in Fig. 10.25 is not there in Fig. 10.26.

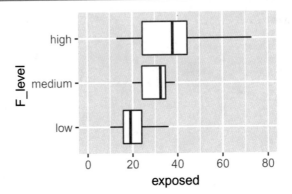

Fig. 10.25 Lead levels in children by father's exposure levels

Fig. 10.26 Lead levels in children in control group by exposure of father of matched treated child

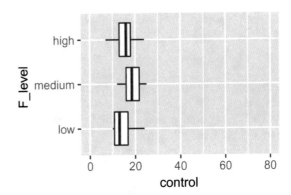

We finally take children whose fathers had high exposure, and within this group, compare children on the basis of father's hygiene.

```
lead$Hyg <- ifelse(lead$hyg == "poor",
    "poor", "ok")

lead %>%
  filter(F_level == "high") %>%
  ggplot(aes(x = Hyg, y = exposed)) +
  geom_boxplot() +
  ylim(0,80) + coord_flip()
```

Figure 10.27 shows that children whose fathers hygiene was worse had higher levels of lead.

10.4.5 Example: Compensation for Injury

We examine the effect of worker compensation laws on the duration a worker was out of work following injury. This study by Meyer, Viscusi and Durbin (1995) is discussed in Rosenbaum (2017). In July 1980, Kentucky raised its maximum benefit

Fig. 10.27 Among children with high exposure fathers, comparison on basis of father's hygiene

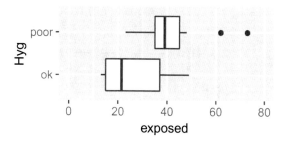

form 131 to 217 dollars per week. This increase only affected workers who were above the earlier limit–high earners.

```
library(wooldridge)
data(injury)
#str(injury)
```

We remove observations with missing values.

```
injury <- injury %>%
  na.omit()
```

We subset the high earners who were in Kentucky.

```
# subset highearners
library(tidyverse)
inj_ky_h <- injury %>%
  filter(ky == 1, highearn == 1)
```

We subset the low earners.

```
inj_ky_l <- injury %>%
  filter(ky == 1, highearn == 0)
```

The variable afchnge is a dummy variable for observations after the policy change. We match the data.

```
## matching for highs, and analysis

match.ky.h <- matchit(afchnge ~ male + married + hosp + indust +
                 injtype + age + lprewage,
             data = inj_ky_h, method = "genetic",
             replace = FALSE, pop.size = 50, print = 0)
#print = 0)
match.ky.h
##
## Call:
## matchit(formula = afchnge ~ male + married + hosp + indust +
##     injtype + age + lprewage, data = inj_ky_h, method = "genetic",
##     replace = FALSE, pop.size = 50, print = 0)
##
```

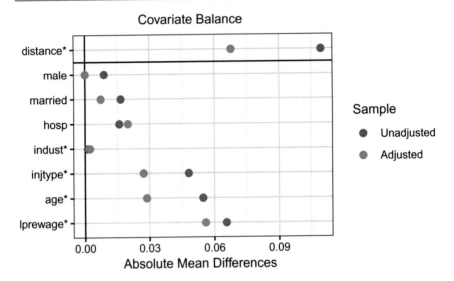

Fig. 10.28 Covariate balance before and after matching

```
## Sample sizes:
##            Control Treated
## All           1128    1103
## Matched       1103    1103
## Unmatched       25       0
## Discarded        0       0
```

All 103 treated observations are matched (Fig. 10.28).

```
love.plot(match.ky.h, stars ="std")
```

```
match_dat_ky_h <- match.data(match.ky.h)
```

We compare the log of duration between after and before groups in the matched data (Fig. 10.29).

```
ggplot(match_dat_ky_h, aes(y = ldurat,
          x = factor(afchnge))) +
  geom_boxplot() +
  coord_flip()
```

```
match_dat_ky_h %>%
  group_by(afchnge) %>%
  summarize(mean_ld = mean(ldurat),
          median_ld = median(ldurat))
## # A tibble: 2 x 3
##    afchnge mean_ld median_ld
##      <int>    <dbl>     <dbl>
```

Fig. 10.29 Log of duration
before and after for high
earners

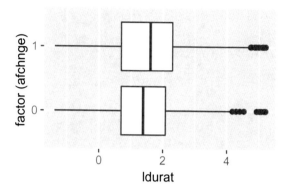

```
## 1        0      1.35       1.39
## 2        1      1.60       1.61
```

The mean of ldurat before was 1.39, after was 1.61.

We use the rbounds package to estimate differences in means in the after and
before group.

```
attach(match_dat_ky_h)

Y <- ldurat
Tr <- afchnge

X <- cbind(male, married,
           hosp, indust,
           injtype, age,
           lprewage)


gen1 <- GenMatch(Tr = Tr, X = X,
                 pop.size = 50,
                 print = 0)

mgen1 <- Match(Y = Y, Tr = Tr, X = X,
               Weight.matrix = gen1,
               replace = FALSE)
summary(mgen1)
##
## Estimate...  0.24846
## SE........  0.048414
## T-stat.....  5.1319
## p.val......  2.8676e-07
##
## Original number of observations..............  2206
## Original number of treated obs..............  1103
## Matched number of observations..............  1103
## Matched number of observations  (unweighted).  1103
```

We get an estimate of 0.25 with a standard error of 0.05.

```
psens(mgen1, Gamma = 1.5, GammaInc = 0.1)
##
##  Rosenbaum Sensitivity Test for Wilcoxon Signed Rank P-Value
##
## Unconfounded estimate .... 0
##
##  Gamma Lower bound Upper bound
##    1.0          0     0.0000
##    1.1          0     0.0002
##    1.2          0     0.0083
##    1.3          0     0.0951
##    1.4          0 .   0.3801
##    1.5          0     0.7355
##
##  Note: Gamma is Odds of Differential Assignment To
##  Treatment Due to Unobserved Factors
##
```

At Gamma = 1.3 the p-value exceeds 0.05 in the sensitivity analysis.

```
hlsens(mgen1, Gamma = 1.5, GammaInc = 0.1)
##
##  Rosenbaum Sensitivity Test for Hodges-Lehmann Point Estimate
##
## Unconfounded estimate ....  0.259
##
##  Gamma Lower bound Upper bound
##    1.0    0.258960     0.25896
##    1.1    0.158960     0.35896
##    1.2    0.058965     0.35896
##    1.3   -0.041035     0.45896
##    1.4   -0.041035     0.55896
##    1.5   -0.041035     0.55896
##
##  Note: Gamma is Odds of Differential Assignment To
##  Treatment Due to Unobserved Factors
##
```

The H-L estimate is 0.25 at Gamma = 0 and the bound crosses zero at Gamma = 1.3.

☞ **Your Turn**   Carry out a similar analysis for the low earners (we have already filtered the data above to get inj_ky_1).

Rosenbaum calls the low earners counterparts; they are not affected by the change in compensation laws, so we can check our results for high earners by seeing whether the duration that low earners stayed out of work increased or not.

10.5   Regression Discontinuity

In a regression discontinuity design, treatment assignment depends on a cutoff value of a variable. We use this knowledge of the treatment assignment to estimate the causal effect.

10.5.1 Simple Example with Synthetic Data

We will do a simulation; so create some synthetic data.

```
library(tidyverse)
set.seed(12)
# sample size
s_size <- 1000
```

The running variable, run, is drawn from a uniform distribution.

```
# running variable
run <- runif(s_size, min = 10,
             max = 50)
```

The treatment variable, treat, is equal to 0 if $run < 20$, else $treat = 1$.

```
# treatment, cutpoint = 20
treat <- ifelse(run < 20,0,1)
```

The outcome is given by: outcome $= 10$ treat $- 0.4$ run $+$ noise

```
# outcome
outcome <-   10 * treat   - 0.4 * run +
  3 * rnorm(s_size)
```

We create a data frame containing the variables.

```
# data frame created
rd_data <- data.frame(
  treat = factor(treat), run, outcome)
```

We plot the data.

```
ggplot(rd_data) +
  geom_point(aes(x = run, y = outcome, shape = treat),
  col = "grey60") + geom_smooth(aes(x = run, y = outcome,
  linetype = treat),
  col = "black") + geom_vline(xintercept = 20)
```

```
## 'geom_smooth()' using method = 'loess' and formula
'y ~ x'
```

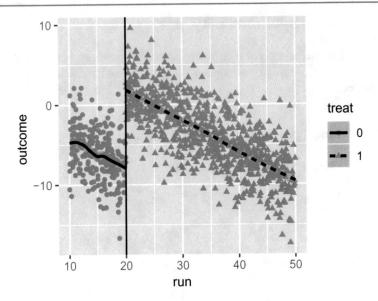

Fig. 10.30 Outcome versus running variable

We see that there is a clear jump at the cutoff point (Fig. 10.30).

Regressing the outcome variable on treatment and the running variable gives us an estimate that is close to the true effect (10).

```
lm(outcome ~ treat + run)
##
## Call:
## lm(formula = outcome ~ treat + run)
##
## Coefficients:
## (Intercept)          treat           run
##     -0.3362         9.8786       -0.3855
```

10.5.2 Example: Minimum Legal Drinking Age (MLDA)

We now work through R code to analyse data relating to the minimum legal drinking age (MLDA) presented in Angrist and Pischke (2015). Did the MLDA of 21 affect death rates in the United States?

```
# read in Stata data
library(foreign)
mlda=read.dta("AEJfigs.dta")
#str(mlda)
```

We have data on `all`, death rates from all causes, and `agecell`, age in months. We create a dummy variable for age over 21.

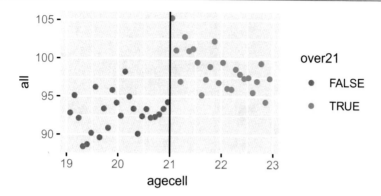

Fig. 10.31 Scatter plot of death rates from all causes versus age in months

```
mlda$over21 = mlda$agecell>=21
```

We plot the data (Fig. 10.31).

```
library(ggplot2)
age3=ggplot(mlda, aes(x = agecell, y = all,colour=over21)) +
  geom_point() +
  geom_vline(xintercept=21)
age3
```

```
## Warning: Removed 2 rows containing missing values
## (geom_point).
```

We add smooths. Irrespective of the type of smooth, we have a clear effect of the minimum legal drinking age (Fig. 10.32).

```
age4=age3 + stat_smooth(method = "lm") +
  stat_smooth(method = "loess")
age4
```

We now use one of the specialized packages in R for regression discontinuity, rddtools (Stigler and Quast 2015).

```
library(rddtools)
```

We remove rows with missing observations.

```
mlda <- mlda %>%
  na.omit()
```

We have to 'declare' the regression discontinuity data.

```
rd_data_2 <- rdd_data(y = all,
                      x=agecell,
                      data=mlda,
```

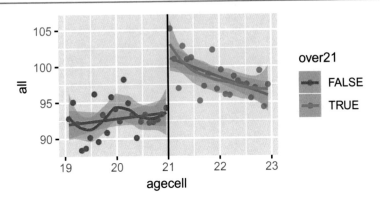

Fig. 10.32 All versus agecell

```
                                    cutpoint=21 )
summary(rd_data_2)
## ### rdd_data object ###
##
## Cutpoint: 21
## Sample size:
##   -Full : 48
##   -Left : 24
##   -Right: 24
## Covariates: no
```

We first use a parametric regression to estimate the treatment effect.

```
reg_para <- rdd_reg_lm(rd_data_2, order=1)
reg_para
## ### RDD regression: parametric ###
##   Polynomial order:  1
##   Slopes:  separate
##   Number of obs: 48 (left: 24, right: 24)
##
##   Coefficient:
##     Estimate Std. Error t value  Pr(>|t|)
## D   7.6627     1.3187   5.8108 6.398e-07 ***
## ---
## Signif. codes:
## 0 '***' 0.001 '**' 0.01 '*' 0.05 '.' 0.1 ' ' 1
```

We get an estimate of 7.7. We plot the parametric regression line (Fig. 10.33).

```
plot(reg_para)
```

☞ **Your Turn**   Change the order of the polynomial in the code above, and
then run.

Fig. 10.33 Parametric regression

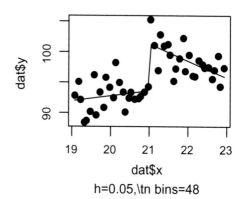

h=0.05,\tn bins=48

Fig. 10.34 Non-parametric regression

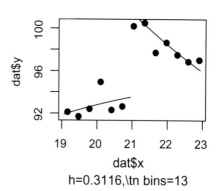

h=0.3116,\tn bins=13

We now use a non-parametric regression; we first get the optimal bandwidth.

```
bw_ik <- rdd_bw_ik(rd_data_2)
bw_ik
##     h_opt
## 1.558202
```

We then estimate and plot the non-parametric regression (Fig. 10.34).

```
reg_nonpara <- rdd_reg_np(rdd_object=rd_data_2, bw=bw_ik)
reg_nonpara
## ### RDD regression: nonparametric local linear###
##   Bandwidth:  1.558202
##   Number of obs: 38 (left: 19, right: 19)
##
##   Coefficient:
##    Estimate Std. Error z value  Pr(>|z|)
## D   9.1894     1.7371  5.2902 1.222e-07 ***
## ---
## Signif. codes:
## 0 '***' 0.001 '**' 0.01 '*' 0.05 '.' 0.1 ' ' 1
plot(reg_nonpara)
```

Fig. 10.35 Placebo test

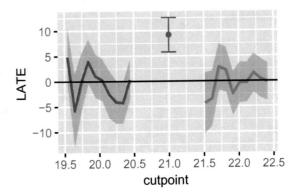

Fig. 10.36 Sensitivity test

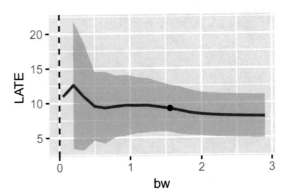

We run placebo and sensitivity tests

```
plotPlacebo(reg_nonpara)
```

Figure 10.35 shows that using cutpoints different from 21 do not give us a statistically significant effect.

```
plotSensi(reg_nonpara, from=0.05, to=3, by=0.15)
```

Figure 10.36 shows that the results are not sensitive to bandwidth.

10.6   Difference-in-Difference

The difference-in-difference method is often used for policy analysis. We first use an example from Wooldridge (2013) to get the basic idea.

10.6.1 Example: Scrap Rate and Training

In this example, the outcome is the scrap rate (how many defective items have to be thrown away) in manufacturing firms in Michigan over 1987 and 1988. The treatment is receipt of a grant for job training.

We can write the scrap rate for 1988 as:

$scrap_{1988} = \beta_0 + \delta_0 1 + \beta_1 grant_{i\,1988} + a_i + u_{i\,1988}$

where i denotes the firm and a_i is a firm-specific factor.
The scrap rate for 1987 is:

$scrap_{1987} = \beta_0 + \beta_1 grant_{i\,1987} + a_i + u_{i\,1987}$

The difference in scrap rates between 1988 and 1987 is:

$scrap_{1988} - scrap_{1987} = \Delta scrap_i = \delta_0 + \beta_1 \Delta grant_i + \Delta u_i$

Differencing helps us remove the confounding effect a_i.
The data is in the wooldridge package.

```
library(wooldridge)
data("jtrain")
```

We remove data for the year 1989.

```
jtrain <- jtrain %>%
  filter(year != 1989)
```

We will use the `plm` package, which handles panel data.

```
library(plm)
jtrain_p <- pdata.frame(jtrain,
  index = c("fcode","year"))
```

We use the `diff()` function to take differences.

```
# Calculate differences
jtrain_p$scrap_d <-
  diff(jtrain_p$scrap)

jtrain_p$grant_d <-
  diff(jtrain_p$grant)
```

We estimate the effect of grant on scrap.

```
mod_did <- lm(scrap_d ~ grant_d,
    data = jtrain_p)
library(texreg)
texreg(list(mod_did), caption = "Dependent variable is scrap",
       caption.above = TRUE)
```

We get a statistically insignificant estimate of -0.74 (Table 10.18).

10.6.2 Simulation

In the previous subsection, we assumed that there was a fixed effect that was a common cause of the outcome and the treatment. Let us call this case A. Differencing removes the fixed effect.

Table 10.18 Dependent variable is scrap

	Model 1
(Intercept)	−0.56
	(0.40)
grant_d	−0.74
	(0.68)
R^2	0.02
Adj. R^2	0.00
Num. obs.	54
RMSE	2.40

$^*p < 0.05$; $^{**}p < 0.01$; $^{***}p < 0.001$

However, what if the initial value of the outcome affects the current value of the outcome and the treatment? Let us call this case B. We may control for the initial value of the outcome.

But we do not know whether the data is generated by a process consistent with case A or case B.

We therefore use simulation to see what happens when we difference or control for initial values in case A and case B.

We generate data for case A.

We have an outcome in period 0 that is a linear function of "fixed" and an error term:

$y_0 = fixed + u_{y0}$

Then in period 1, we have a treatment that is determined by fixed:

$treat = 1, if : fixed < 0, else = 0$

Treatment has an effect eff, that affects the outcome in addition to fixed.

$y_1 = fixed + eff * treat + u_{y1}$

We generate the synthetic data with the following code:

```
ss <- 3000 # sample size
eff <- 3 # effect is 3
fixed <- rnorm(ss) # random normal
treat <- ifelse(fixed < 0,1,0)
uy1 <- rnorm(ss)
uy0 <- rnorm(ss)
y1 <- fixed + eff * treat + uy1
y0 <- fixed + uy0
```

In m_d1, we estimate the difference of means. In m_d2 we control for the initial value of y, y0. In m_d3, we use difference-in-difference.

```
library(texreg)
# difference of means
m_d1 <- lm(y1 ~ treat)
# controlling for y0
m_d2 <- lm(y1 ~ y0 + treat)
# diff in diff
```

Table 10.19 Case A results (True effect of treat is 3)

	Model 1	Model 2	Model 3
(Intercept)	0.78***	0.57***	−0.00
	(0.03)	(0.03)	(0.04)
treat	1.40***	1.81***	2.92***
	(0.04)	(0.05)	(0.05)
y0		0.27***	
		(0.02)	
R^2	0.26	0.32	0.52
Adj. R^2	0.26	0.32	0.52
Num. obs.	3000	3000	3000
RMSE	1.17	1.12	1.41

$^{*}p < 0.05; {^{**}}p < 0.01; {^{***}}p < 0.001$

```
m_d3 <- lm(I(y1 - y0) ~ treat)
texreg(list(m_d1, m_d2, m_d3), caption = "Case A results",
       caption.above = TRUE)
```

In Table 10.19, Model 3 (difference-in-difference) comes close to the true effect. Model 2 (controlling for initial value) gives us an underestimate, and Model 1 (difference of means) is further from the true effect.

We generate data for case B.

We have an outcome in period 0 that is a random variable:

$y_0 = u_{y0}$

The initial outcome influences treatment:

$treat = 1, if : y_0 < 2.5, else = 0$

In period 1, treatment has an effect eff, that affects the outcome in addition to the value of the initial outcome.

$y_1 = beta * y_0 + eff * treat + u_{y1}$

We generate the synthetic data with the following code:

```
ss <- 3000
eff <- 3
y0 <- runif(ss, min = 1, max = 4)
treat <- ifelse(y0 < 2.5,1,0)
uy1 <- rnorm(ss)
uy0 <- rnorm(ss)
y1 <- 0.3 * y0 + eff * treat + uy1


# difference of means
m_d4 <- lm(y1 ~ treat)
# controlling for y0
m_d5 <- lm(y1 ~ y0 + treat)
# diff in diff
m_d6 <- lm(I(y1 - y0) ~ treat)
texreg(list(m_d4, m_d5, m_d6), caption = "Case B results",
       caption.above = TRUE)
```

Table 10.20 Case B results (True effect of treat is 3)

	Model 1	Model 2	Model 3
(Intercept)	0.98***	0.13	−2.27***
	(0.03)	(0.14)	(0.03)
treat	2.55***	2.94***	4.04***
	(0.04)	(0.07)	(0.04)
y0		0.26***	
		(0.04)	
R^2	0.61	0.62	0.79
Adj. R^2	0.61	0.62	0.79
Num. obs.	3000	3000	3000
RMSE	1.01	1.00	1.05

$^* p < 0.05; ^{**} p < 0.01; ^{***} p < 0.001$

In Table 10.20 we see that the difference-in-difference gives us an overestimate, model 2 (controlling for initial value) is close and the simple difference in means is a large underestimate.

So we see that the difference-in-difference has a crucial assumption, that the treatment and control group have parallel trends. This can be graphically examined.

Hence, like in regression discontinuity, graphical examination can help us judge the validity of difference-in-difference in a given context.

10.6.3 Example: Banks in Business

In this example from Angrist and Pischke (2015) we examine the role of monetary policy, during the Great Depression in different districts of the US Federal Reserve system. We compare the effect of easy lending to troubled banks as practised by the Atlanta Fed, that ran the Sixth District, to the restrictive policy of the St. Louis Fed that ran the Eighth District. The outcome variable is the number of banks in business. The treatment group is the sixth district, and the control group is the eighth district.

We get the data:

```
library(readr)
banks <- read_csv("banks.csv")


## Parsed with column specification:
## cols(
##   date = col_double(),
##   weekday = col_character(),
##   day = col_double(),
##   month = col_double(),
##   year = col_double(),
##   bib6 = col_double(),
##   bio6 = col_double(),
##   bib8 = col_double(),
```

```
##   bio8 = col_double()
## )
#str(banks)
```

We calculate the mean number of banks in business each year in the 6th and 8th districts.

```
library(tidyverse)
bankag <- banks %>%
  group_by(year) %>%
  summarize(bib6m=mean(bib6),
            bib8m=mean(bib8))
head(bankag)
## # A tibble: 6 x 3
##     year bib6m bib8m
##    <dbl> <dbl> <dbl>
## 1   1929   141  170.
## 2   1930  136.  165.
## 3   1931  120.  132.
## 4   1932  113.  120.
## 5   1933  105.  112.
## 6   1934   102  110.
```

We stack, or gather, the banks in business for the sixth and eighth districts in one column.

```
bankag2 <- gather(bankag,
          "bty","num",2:3)
```

We filter for years 1930 and 1931.

```
bankag3 <- filter(bankag2,
   year == 1930 | year == 1931)
bankag3
## # A tibble: 4 x 3
##     year bty       num
##    <dbl> <chr>   <dbl>
## 1   1930 bib6m   136.
## 2   1931 bib6m   120.
## 3   1930 bib8m   165.
## 4   1931 bib8m   132.
```

We plot the banks in business in 1930 and 1931 (Fig. 10.37). Our difference-in-difference estimate is $= (120 - 136) - (132 - 165) = -16 - (-33) = 17$.

```
ggplot(bankag3, aes(
  x=year,
  y=num,
  colour=bty)) +
  geom_line()
```

We can explore the parallel trends assumption; Fig. 10.38 shows that the trends are indeed parallel before 1930 and after 1931.

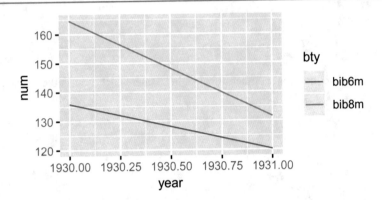

Fig. 10.37 Difference-in-difference for 1930 and 1931

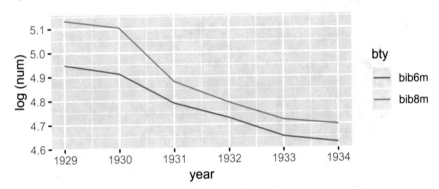

Fig. 10.38 Parallel trends assumption is supported by graph

```
ggplot(bankag2, aes(
  x=year,
  y=log(num),
  colour=bty)) +
  geom_line()
```

10.7   Example: Manski Bounds for Crime and Laws

Manski and Pepper (2018) study the effect of right to carry laws on crime. We will present a simplified version of their paper.

The state of Virginia allowed guns to be carried (RTC was implemented) in 1989. Maryland in contrast did not do so in the period of the study (1970–2007).

Fig. 10.39 Murder rates in
Virginia (solid line) and
Maryland (dashed line).
Virginia enacted a right to
carry statute in 1989 (vertical
dotted line)

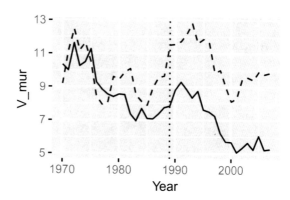

We load a portion of the data that they used: murder rates in Virginia and Maryland
by year.

```
Manski <- read_csv("Manski.csv")
```

```
## Parsed with column specification:
## cols(
##   Year = col_double(),
##   V_mur = col_double(),
##   M_mur = col_double()
## )
```

We graph the data relating to murder rates (Fig. 10.39).

```
ggplot(Manski) +
  geom_line(aes(x = Year,
  y = V_mur), linetype = "solid") +
  geom_line(aes(x = Year,
  y = M_mur), linetype = "dashed") +
  geom_vline(xintercept = 1989,
             linetype = "dotted")
```

Except for a few years in the 1970s, murder rates in Maryland were higher, often a
lot higher, than those in Virginia. Murder rates rose in 1980s and then declined sharply
in the mid-1990s (Fig. 10.39). We can see that the murder rates in Virginia before
the RTC vary considerably. The murder rates in Virginia are not moving parallel to
the murder rates in Maryland. This makes difference-in-difference implausible.

We represent the crime outcome (murder) by Y. Using VA to denote Virginia, and
using (1) to denote treatment, we can represent the potential outcome of the murder
rate in the year 1990 in Virginia by

$$Y_{VA,1990}(1).$$

Since Virginia passed its right to carry statute (RTC) in 1989, the potential outcome
of crime in Virginia in 1990 given the RTC is observed. However, the potential
outcome of crime in Virginia in 1990 in the absence of RTC is not observed. Using
(0) to denote absence of treatment, the treatment effect is given by:

Fig. 10.40 Difference in murder rates in Virginia and Maryland between 1970 and 1988

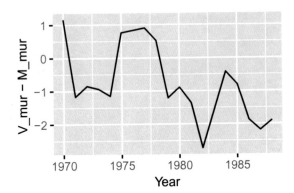

$Y_{VA,1990}(1) - Y_{VA,1990}(0)$.

Given that the usual difference-in-difference assumptions are implausible, Manski and Pepper (2018) impose assumptions that relate counterfactual quantities to observed data, and give us bounded, not point, estimates.

10.7.1 Bounds with Maryland as Counterfactual

Since $Y_{VA,1990}(0)$ is not observed, we can use comparison with Maryland, i.e. $Y_{MD,1990}(0)$ to fill in the value assuming:

$Y_{VA,1990}(0) - Y_{MD,1990}(0) = 0$

We don't observe $Y_{VA,1990}(0)$, so can't know its validity directly.

However, from 1970 to 1988, we can observe:

$Y_{VA,d}(0) - Y_{MD,d}(0) = \delta_{VAMD,d}$

```
ggplot(Manski[1:19,]) +
  geom_line(aes(x = Year,
      y = V_mur - M_mur))
```

The difference, $Y_{VA,d}(0) - Y_{MD,d}(0) = \delta_{VAMD,d}$ goes from a high of 1 to a low of -2.7. So from 1990 onwards we can use $Y_{MD,d}(0)$ to fill in for $Y_{VA,d}(0)$, but raising it by 1 for one bound of the treatment effect and lowering it by 2.7 for another bound of the treatment effect (Fig. 10.40).

```
ggplot(Manski[20:38,]) +
  geom_line(aes(x = Year,
              y = V_mur),
          linetype = "solid") +
  geom_line(aes(x = Year,
              y = M_mur),
          linetype = "dashed") +
  geom_line(aes(x = Year,
              y = M_mur + 1),
          linetype = "dotted") +
  geom_line(aes(x = Year,
```

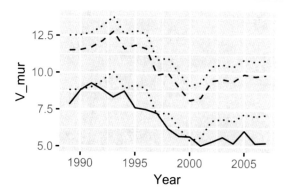

Fig. 10.41 Solid line is murder rates in Virginia, dashed line is murder rates in Maryland, upper dotted line is Maryland rates +1, lower dotted line is Maryland rates −2.7

Fig. 10.42 Treatment effect using Maryland crime rates +1 and Maryland crime rate −2.7 as counterfactuals

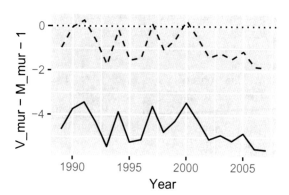

```
        y = M_mur - 2.7),
   linetype = "dotted")
```

Figure 10.41 shows murder rates in Virginia (solid line), murder rates in Maryland (dashed line); the upper dotted line represents Maryland rates +1, lower dotted line represents Maryland rates −2.7.

```
ggplot(Manski[20:38,]) +
  geom_line(aes(x = Year,
            y = V_mur - M_mur - 1),
          linetype = "solid") +
  geom_line(aes(x = Year,
            y = V_mur - M_mur + 2.7),
          linetype = "dashed") +
  geom_hline(yintercept = 0,
          linetype = "dotted")
```

Figure 10.42 shows the lower and upper bounds of the estimated treatment effects.

Fig. 10.43 Historical delta

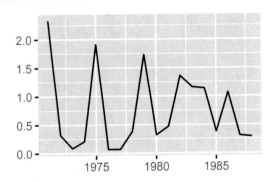

10.7.2 Bounds Based on Difference-in-Difference

Since $Y_{VA,1990}(0)$ is not observed, we can use a bounded version of the usual difference-in-difference (DID) assumption.

$$I_{diff} = [Y_{VA,1990}(0) - Y_{VA,1989}(0)] - [Y_{MD,1990}(0) - Y_{MD,1989}(0)] = 0$$

where I_{diff}, if zero, is a difference that leads to identification of the effect of treatment.

We don't observe $Y_{VA,1990}(0)$, so can't know its validity directly.

However, from 1970 to 1988, we can observe:

$$|[Y_{VA,d+1}(0) - Y_{VA,d}(0)] - [Y_{MD,d+1}(0) - Y_{MD,d}(0)]| = \delta_{d+1}$$

We will now calculate this and plot it. It will be convenient to use the ts object for time series (Fig. 10.43).

```
Manski_ts <- ts(Manski[2:3],
  start = 1970, end = 2007)
Manski_ts2 <- window(Manski_ts,
    start = 1970, end = 1988)
head(Manski_ts2)
##       V_mur M_mur
## [1,] 10.34  9.18
## [2,]  9.99 11.16
## [3,] 11.62 12.47
## [4,] 10.25 11.19
## [5,] 10.49 11.64
## [6,] 11.27 10.51
```

```
Manski_ts2_d <- diff(Manski_ts2)

diff_V_M <- Manski_ts2_d[,1] -
  Manski_ts2_d[,2]

library(ggfortify)

autoplot(abs(diff_V_M))
```

From the series of δ_{d+1}, we could pick a suitable value, one perhaps is the 0.75 quantile of the series, which we denote as $\delta_{0.75}$, (delta_0.75) = 1.155.

```
quantile(abs(diff_V_M),
        probs = c(0,0.25,0.5,0.75,1))
##    0%   25%   50%   75%  100%
## 0.070 0.295 0.380 1.155 2.330
delta_0.75 <- quantile(abs(diff_V_M),
        probs = c(0.75))
delta_0.75
##   75%
## 1.155
```

Since the DID assumption does not hold in our data from 1970 to 1988, we can use a bound to study the treatment effect for Virginia for the effect of the RTC from 1990 onwards:

$$|[Y_{VA,d}(0) - Y_{VA,1988}(0)] - [Y_{MD,d}(0) - Y_{MD,1988}(0)]| =< \delta_{0.75}$$

The code below does the calculations.

```
Manski_ts3 <- window(Manski_ts,
      start = 1990, end = 2007)

Manski_ts3
## Time Series:
## Start = 1990
## End = 2007
## Frequency = 1
##       V_mur M_mur
## 1990   8.81 11.55
## 1991   9.28 11.72
## 1992   8.84 12.16
## 1993   8.34 12.79
## 1994   8.74 11.61
## 1995   7.62 11.86
## 1996   7.50 11.63
## 1997   7.25  9.86
## 1998   6.22 10.00
## 1999   5.70  8.99
## 2000   5.67  8.12
## 2001   5.06  8.30
## 2002   5.33  9.44
## 2003   5.64  9.56
## 2004   5.23  9.41
## 2005   6.06  9.90
## 2006   5.22  9.75
## 2007   5.26  9.84
window(Manski_ts, start = 1988, end = 1988)
## Time Series:
## Start = 1988
## End = 1988
## Frequency = 1
##       V_mur M_mur
## 1988   7.75  9.64
TE_strong <- Manski_ts3[,1] - 7.75 -
  Manski_ts3[,2] + 9.64
```

Table 10.21 Treatment effect with strong difference-in-difference assumption, and lower and upper bound DID assumption

	Year	TE_lb	TE_strong	TE_ub
1	1990	−2.00	−0.80	0.30
2	1991	−1.70	−0.60	0.60
3	1992	−2.60	−1.40	−0.30
4	1993	−3.70	−2.60	−1.40
5	1994	−2.10	−1.00	0.20
6	1995	−3.50	−2.30	−1.20
7	1996	−3.40	−2.20	−1.10
8	1997	−1.90	−0.70	0.40
9	1998	−3.00	−1.90	−0.70
10	1999	−2.60	−1.40	−0.20
11	2000	−1.70	−0.60	0.60
12	2001	−2.50	−1.40	−0.20
13	2002	−3.40	−2.20	−1.10
14	2003	−3.20	−2.00	−0.90
15	2004	−3.40	−2.30	−1.10
16	2005	−3.10	−1.90	−0.80
17	2006	−3.80	−2.60	−1.50
18	2007	−3.80	−2.70	−1.50

```
TE_lb <- TE_strong - delta_0.75

TE_ub <- TE_strong + delta_0.75

Manski_eff <- cbind(TE_lb, TE_strong, TE_ub)

Manski_eff <- round(Manski_eff,1)

Manski_eff_d <- data.frame(Year = 1990:2007, Manski_eff)
```

```
library(xtable)
teff <- xtable(Manski_eff_d,
     caption = "Treatment effect with strong difference-
     in-difference assumption, and lower and upper bound DID
     assumption.")
print(teff, caption.placement = "top")
```

Table 10.21 displays the results using the `xtable` package (Dahl et al. 2019); Fig. 10.44 is a plot.

```
autoplot(Manski_eff, facets = F) +
  geom_hline(yintercept = 0,
             linetype = "dotted")
```

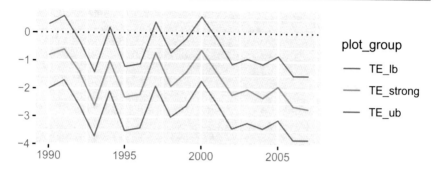

Fig. 10.44 Treatment effect with strong difference-in-difference assumption, and lower and upper bound DID assumption

Figures 10.42 and 10.43 employ different data-based assumptions about counterfactual values. In both figures, we get a general reduction in murder rates due to the RTC, with effects varying over years. Manski and Pepper (2018) find that the treatment effects vary by type of crime and over time.

10.8   Instrumental Variables

If $y = \beta_0 + \beta_1 x + e$ and $cov(x, e) \neq 0$ then a regression of y on x will not give us an unbiased estimate of $\beta 1$.

However, if there is a variable z that affects x, in such a way that $cov(z, x) \neq 0$ and $cov(z, e) = 0$ then we can use z as an instrumental variable.

$\beta_1 = cov(z, y)/cov(z, x)$ in the simple case above.

10.8.1   Simulation

We simulate data where:

$U \rightarrow X, U \rightarrow Y$; IV is needed if U is an unobservable variable.

$X \rightarrow Y$

$Z \rightarrow X$

The sample size and the strength of the effect of Z on X play a key role in IV. We generate the data (note $\beta_X = 1$):

```
library(tidyverse)
library(AER)
sample_size = 300
coef_Z = 0.9
viol = 0

    Z <- runif(sample_size,
            min = 1, max = 5)
    U <- runif(sample_size,
            min = 1, max = 5) + viol*Z
```

Table 10.22 OLS results for Y (True effect of X is 1)

	Model 1	Model 2
(Intercept)	0.79**	−0.06
	(0.26)	(0.20)
X	1.40***	0.98***
	(0.04)	(0.04)
U		1.05***
		(0.07)
R^2	0.77	0.88
Adj. R^2	0.77	0.88
Num. obs.	300	300
RMSE	1.37	1.02

$^*p < 0.05; ^{**}p < 0.01; ^{***}p < 0.001$

Table 10.23 IV results for Y (True effect of X is 1)

	Model 1
(Intercept)	3.21***
	(0.52)
X	0.98***
	(0.09)
R^2	0.70
Adj. R^2	0.70
Num. obs.	300
RMSE	1.57

$^*p < 0.05; ^{**}p < 0.01; ^{***}p < 0.001$

```
X <- U + rnorm(sample_size) + coef_Z *Z
Y <- U + X + rnorm(sample_size)
```

```
mod1OLS <- lm(Y ~ X)
mod2OLS <- lm(Y ~ X + U)
texreg(list(mod1OLS, mod2OLS), caption = "OLS results for Y",
       caption.above = TRUE)
```

Table 10.22 shows that in Model 1, with U not observed, we get a biased, statistically significant, estimate of β_X. If we observed U, then controlling for it gives an unbiased estimate of β_X (Model 2).

We now use instrumental variables regression.

```
library(AER)
ModIV <- ivreg(Y ~ X | Z)
texreg(list(ModIV), caption = "IV results for Y",
       caption.above = TRUE)
```

Table 10.23 shows that instrumental variable (IV) regression gives us an unbiased estimate of β_X.

We now create a function to carry out IV simulations.

```
IVsamD <- function(sample_size,
    coef_Z,viol = 0) {
num_loops = 300
#sample_size = 30
#coef_Z = 0.5

OLS1 <- numeric(num_loops)
OLS2 <- numeric(num_loops)
IV <- numeric(num_loops)
for (i in 1: num_loops) {

  U <- runif(sample_size,
            min = 1, max = 5)
  Uy <- rnorm(sample_size)
  Z <- runif(sample_size,
            min = 1, max = 5) + viol*Uy
  X <- U + rnorm(sample_size) + coef_Z *Z
  Y <- U + X + Uy
  OLS1[i] <- summary(lm(Y ~ X))$coef[2]
  OLS2[i] <- summary(lm(Y ~ X + U))$coef[2]
  IV[i] <- summary(ivreg(Y ~ X | Z))$coef[2]
}
reg_IV <- tibble(OLS1,OLS2,IV)
reg_IV
library(tidyr)
reg_IV_s <- reg_IV %>%
  gather(Estimator,value,OLS1:IV)
reg_IV_s
library(ggplot2)
ggplot(reg_IV_s, aes(value,
                    colour = Estimator)) +
  geom_density() +
  xlim(c(-1,2)) +
  geom_vline(xintercept = 1, lty = 2)
}
```

We now use the IV function.

```
IVsamD(sample_size = 30,
    coef_Z = 1, viol = 0)
```

The IV sampling distribution has greater spread though consistent (Fig. 10.45).

☞ **Your Turn** Copy the code for the IV simulation function. Try the following scenarios (they are written as R code comments; you have to remove the hash):

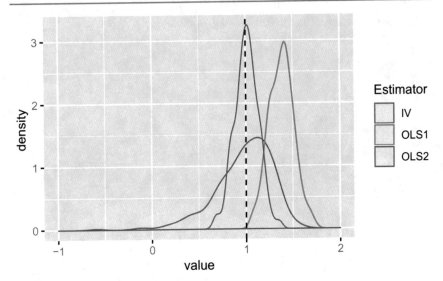

Fig. 10.45 Sampling distributions of estimators of X on Y. True effect is 1. IV uses Z as instrument, OLS1 is Y regressed only on X, OLS2 is Y regressed on X and U

```
# Large sample size
#IVsamD(sample_size = 300, coef_Z = 1, viol = 0)
# weak instrument
#IVsamD(sample_size = 300, coef_Z = 0.2, viol = 0)
# exclusion restriction violated
#IVsamD(sample_size = 300, coef_Z = 1, viol = 0.5)
```

10.8.2 Example: Demand for Cigarettes

Public policy often aims to reduce smoking because of its effects on health. If taxes are used to restrict smoking, the question arises–how will consumption be affected? This example is from Stock and Watson (2011).

We will work with data on cigarettes in the AER package, and rename it for convenience. The data is for the states in the US, for the years 1985 and 1995.

```
library(AER)
data("CigarettesSW", package = "AER")
# rename dataframe for convenience
Cig <- CigarettesSW
```

We have data on annual per capita cigarette sales in packs, packs, and price, price, and the consumer price index, cpi (Fig. 10.46).

Fig. 10.46 Log prices versus log packs

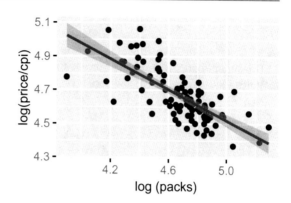

```
library(ggplot2)
ggplot(Cig, aes(x=log(packs),
            y=log(price/cpi))) +
  geom_point() +
  stat_smooth(method = "lm")
```

We will be removing fixed effects by differencing, and will use instrumental variables estimation on the differenced data.

We create new variables with the mutate function. Below we convert our price, income and tax variables to real terms.

```
Cig <-
  Cig %>%
  mutate(rprice = price/cpi)

Cig <-
  Cig %>%
  mutate(rincome =
      income/population/cpi)
# taxs is sales tax, tax is tax
Cig <-
  Cig %>%
  mutate(rtaxs=
          (taxs-tax)/cpi)

Cig <-
  Cig %>%
  mutate(rtax=
          tax/cpi)
```

We create separate data frames for the years 1985 and 1995 with the data verb filter.

```
Cig85 <-
  Cig %>%
  filter(year==1985)
Cig95 <-
```

```
Cig %>%
filter(year==1995)
```

Transmute works like mutate but does not add to the other data. We pull out variables for 1985 and 1995, and then take the difference.

```
pack_85  <-
  Cig85 %>%
  transmute(pack_85 = log(packs))

pack_95 <-
  Cig95 %>%
  transmute(pack_95=log(packs))

# Difference 1995 and 1985
pack_diff  <- pack_95$pack_95 -
  pack_85$pack_85
```

In the last line we took a long-term difference in the number of cigarette packs between 1995 and 1985; we will be studying the long-term elasticity.

We carry out calculations for price:

```
rprice85  <-
  Cig85 %>%
  transmute(rprice85 = log(rprice))

rprice95  <-
  Cig95 %>%
  transmute(rprice95 = log(rprice))

rpricediff  <- rprice95$rprice95 -
  rprice85$rprice85
```

For income and taxes

```
i85  <-
  Cig85 %>%
  transmute(i85 = log(rincome))

i95  <-
  Cig95 %>%
  transmute(i95 = log(rincome))

idiff  <- i95$i95 - i85$i85

ts85  <-
  Cig85 %>%
  transmute(ts85 = rtaxs)

ts95  <-
  Cig95 %>%
  transmute(ts95 = rtaxs)
```

```
tsdiff  <- ts95$ts95 - ts85$ts85


t85  <-
  Cig85 %>%
  transmute(t85 = rtax)

t95  <-
  Cig95 %>%
  transmute(t95 = rtax)

tdiff  <- t95$t95 - t85$t85
```

The dependent variable is quantity of cigarettes, which is regressed on price (endogenous) and income (exogenous); the instrument is general sales taxes. All variables are differences between 1995 and 1985.

```
# convenience function: HC1 covariances
library(estimatr)
hc1 <- function(x) vcovHC(x, type = "HC1")
mod1 <- iv_robust(pack_diff ~ rpricediff +
        idiff| idiff + tsdiff,
        diagnostics = TRUE)
mod2 <- iv_robust(pack_diff ~ rpricediff +
        idiff| idiff + tdiff)
mod3 <- iv_robust(pack_diff ~ rpricediff +
        idiff| idiff + tsdiff
        + tdiff, diagnostics = TRUE)



library(texreg)
texreg(list(mod1, mod2, mod3),
        caption = "Instrumental variable estimates",
        caption.above = TRUE)
```

The estimates are in Table 10.24. We conduct the overidentification test for validity of instruments; in Model 3 we have two instruments.

```
summary(mod3)$diagnostic_overid_test
##      value         df    p.value
## 4.08518901 1.00000000 0.04326061
```

The null hypothesis that both instruments are exogenous is rejected at the 5 percent level. Stock and Watson (2011, p. 448) make a case for Model 1. They argue: 'We think that the case for the exogeneity of the general sales tax is stronger than that for the cigarette-specific tax, because the political process can link changes in the cigarette-specific tax to changes in the cigarette market and smoking policy'.

The F-statistic for Model 1 is fine.

Table 10.24 Instrumental variable estimates

	Model 1	Model 2	Model 3
(Intercept)	−0.12	−0.02	−0.05
	[−0.26; 0.02]	[−0.16; 0.13]	[−0.18; 0.08]
rpricediff	−0.94*	−1.34*	−1.20*
	[−1.37; −0.51]	[−1.84; −0.85]	[−1.62; −0.78]
idiff	0.53	0.43	0.46
	[−0.18; 1.23]	[−0.19; 1.05]	[−0.18; 1.10]
R^2	0.55	0.52	0.55
Adj. R^2	0.53	0.50	0.53
Num. obs.	48	48	48
RMSE	0.09	0.09	0.09

* 0 outside the confidence interval

```
summary(mod1)$diagnostic_first_stage_fstatistic
##          value          nomdf          dendf
## 3.201544e+01  1.000000e+00  4.500000e+01
##        p.value
## 1.001477e-06
```

The long-run elasticity is about −0.9, which is somewhat elastic.

10.9   Resources

For Better Understanding

For understanding experiments and the potential outcomes approach, Rosenbaum (2017), for understanding causal graphs, Pearl et al. (2016), for programme evaluation, Josselin and Le Maux (2017), for econometric methods for causal inference explained with great humour, Angrist and Pischke (2015).

For Going Further

Morgan and Winship (2014), Abadie and Catteneo (2018), Manski and Pepper (2018).

References

Abadie, A., and M.D. Catteneo. 2018. Econometric methods for program evaluation. *Annual Review of Economics* 10: 465–503.

Angrist, J.D., and J. Pischke. 2015. *Mastering 'metrics - The path from cause to effect*. Princeton: Princeton University Press.

Chattopadhyay, R., and E. Duflo. 2004. Women as policy makers: Evidence from a randomized policy experiment in India. *Econometrica* 72 (5): 1409–1443.

Coppock, A. 2019. ri2: Randomization inference for randomized experiments. R package version 0.1.2. https://CRAN.R-project.org/package=ri2.

Dahl, D.B., D. Scott, C. Roosen, A. Magnusson, and J. Swinton. 2019. xtable: Export Tables to LaTeX or HTML. R package version 1.8-4. https://CRAN.R-project.org/package=xtable.

Dehejia, R.H., and S. Wahba. 1999. Causal effects in nonexperimental studies: Reevaluating the evaluation of training programs. *Journal of the American Statistical Association* 94 (448): 1053–1062.

Elwert, F. 2013. Graphical causal models. In *Handbook of causal analysis for social research*, ed. S.L. Morgan, 245–274. New York: Springer.

Freedman, D.A. 1983. A note on screening regression equations. *The American Statistician* 37 (2): 152–155.

Gelman, A., and J. Hill. 2007. *Data analysis using regression and multilevel/hierarchical models (Analytical methods for social research)*. Cambridge: Cambridge University Press.

Greifer, N. 2019. cobalt: Covariate balance tables and plots. R package version 3.8.0. https://CRAN.R-project.org/package=cobalt.

Hill, R.C., W.E. Griffiths, and G.C. Lim. 2018. *Principles of econometrics*. New York: Wiley.

Ho, D.E., K. Imai, G. King, and E.A. Stuart. 2011. MatchIt: Nonparametric preprocessing for parametric causal inference. *Journal of Statistical Software* 42 (8): 1–28. https://www.jstatsoft.org/v42/i08/.

Imai, K. 2018. *Quantitative social science - An introduction*. Princeton: Princeton University Press.

Josselin, J.-M., and B. Le Maux. 2017. *Statistical tools for program evaluation: Methods and applications to economic policy, public health, and education*. Berlin: Springer.

Kahneman, D. 2011. *Thinking, fast and slow*. London: Penguin Books.

Keele, L.J. 2014. rbounds: Perform Rosenbaum bounds sensitivity tests for matched and unmatched data. R package version 2.1. https://CRAN.R-project.org/package=rbounds.

Lalonde, R.J. 1986. Evaluating the econometric evaluations of training programs with experimental data. *The American Economic Review* 76 (4): 604–620.

Leifeld, P. 2013. texreg: Conversion of statistical model output in R to LaTeX and HTML tables. *Journal of Statistical Software* 55 (8): 1–24. http://www.jstatsoft.org/v55/i08/.

Manski, C.F., and J.V. Pepper. 2018. How do right-to-carry laws affect crime rates? Coping with ambiguity using bounded variation assumptions. *Review of Economics and Statistics* 100 (2): 232–244.

Meyer, B.D., W.K. Viscusi, and D.L. Durbin. 1995. Workers' compensation and injury duration: Evidence from a natural experiment. *The American Economic Review* 85 (3): 322–340.

Morgan, S.L., and C. Winship. 2014. *Counterfactuals and causal inference: Methods and principles for social research (Analytical methods for social research)*. Cambridge: Cambridge University Press.

Pearl, J., M. Glymour, and N.P. Jewell. 2016. *Causal inference in statistics: A primer*. New York: Wiley.

Rosenbaum, P. 2005. Sensitivity analysis in observational studies. In *Encyclopedia of statistics in behavioural science*, ed. B.S. Everitt, D.C. Howell, 1809–1814. New York: Wiley.

Rosenbaum, P. 2017. *Observation and experiment - An introduction to causal inference*. London: Harvard University Press.

Rubin, D.B. 2008. Statistical inference for causal effects, with emphasis on applications in epidemiology and medical statistics. In *Handbook of Statistics*, vol. 27, ed. C.R. Rao, J.P. Miller, D.C. Rao. 2008. Amsterdam: Elsevier.

Sekhon, J.S. 2011. Multivariate and propensity score matching software with automated balance optimization: The matching package for R.

Stigler, M., and B. Quast. 2015. rddtools: Toolbox for Regression Discontinuity Design ('RDD'). R package version 0.4.0. https://CRAN.R-project.org/package=rddtools.

Stock, J.H., and M.W. Watson. 2011. *Introduction to econometrics*. Boston: Addison-Wesley.

Wooldridge, J. 2013. *Introductory econometrics: A modern approach*. Delhi: Cengage.

Part V
Accessing, Analysing and Interpreting Growth Data

Growth Data and Models

11

11.1 Introduction

Growth is a key economic policy issue. Here we will follow the intuitive treatment of growth provided by Jones (2018), who builds on an interplay of data and models. The stages in the development of the ideas presented by him are:

1. A look at the very long run growth data of the Maddison project.
2. Development of a simple production model to understand differences in per capita GDP across countries, and then viewing the Penn World Tables data through the lens of this model.
3. Simulating the Solow model of economic growth and understanding its properties.
4. Simulate a very simplified version of Romer's model and see how that helps us to incorporate the concept of ideas in growth.
5. In addition, we use the World Development Indicators data to visualize growth in recent decades.

11.2 Example: Growth

We install the Maddison package (Persson 2015) and then load it, along with tidyverse (Wickham 2017).

```
library(tidyverse)
library(maddison)
str(maddison)
## Classes 'tbl_df', 'tbl' and 'data.frame':    45318 obs. of
    9 variables:
```

Fig. 11.1 US growth over
time

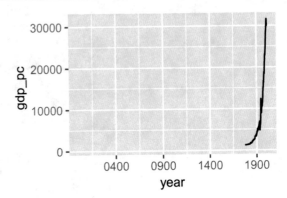

```
##   $ year            : Date, format:   ...
##   $ country_original: chr  "Austria" "Austria" "Austria"
     "Austria" ...
##   $ gdp_pc          : num  NA NA NA NA NA NA NA NA NA NA ...
##   $ country         : chr  "Austria" "Austria" "Austria"
     "Austria" ...
##   $ iso2c           : chr  "AT" "AT" "AT" "AT" ...
##   $ iso3c           : chr  "AUT" "AUT" "AUT" "AUT" ...
##   $ continent       : chr  "Europe" "Europe" "Europe"
     "Europe" ...
##   $ region          : chr  "Western Europe" "Western Europe"
     "Western Europe" "Western Europe" ...
##   $ aggregate       : logi  FALSE FALSE FALSE FALSE FALSE
     FALSE ...
```

We first look at the values for the United States, so we filter the data. As always,
the data wrangling and the plotting work together.

```
mad1 <- maddison %>%
  filter(iso2c == "US")
ggplot(mad1, aes(x = year, y = gdp_pc)) +
  geom_line()
```

```
## Warning: Removed 33 rows containing missing values
(geom_path).
```

Figure 11.1 gives us a sense of the data, but it appears that it is best to look at the
values after 1800.

```
mad2 <- maddison %>%
  filter(iso2c == "US" & year>=
          as.Date("1800-01-01"))
```

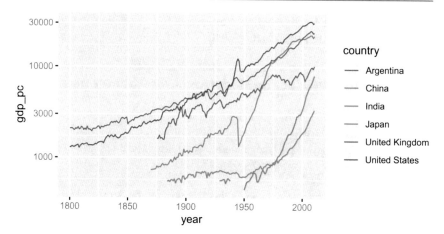

Fig. 11.2 Very long run growth in select countries

☞ **Your Turn** Plot the GDP per capita in the USA since 1800, using the filtered tibble mad2. The x variable is year, y variable is gdp_pc. After this plot, plot GDP per capita with log scale for the y-axis (scale_y_log10()).

We now plot data for a few select countries. This provides a rich historical comparison.

```
mad3 <- maddison %>%
  filter(country %in% c("India", "China",
    "Argentina", "Japan", "United States",
    "United Kingdom") &
      year>=
      as.Date("1800-01-01"))

ggplot(mad3, aes(x = year, y = gdp_pc,
                 colour = country)) +
  geom_line()+
  scale_y_log10()
```

```
## Warning: Removed 90 rows containing missing values
(geom_path).
```

Several features of long run growth are evident in Fig. 11.2. The US and the UK have experienced steady growth over a long period. The US got richer than the UK over time. After the Second World War Japan got a shock, but then grew rapidly to catch up with the US and the UK. China's surge has captured the world's imagination. In recent decades, India too has grown fast. Argentina, though initially relatively prosperous, fell away in its growth.

11.3   Example: Production Model and Crosscountry Data

Jones exposits a production model and takes it to data.

Let production, Y be given by: $Y = AK^a L^{1-a}$

where A is a production parameter, and K and L are labour and capital.

Dividing through by L, we get: $Y/L = AK^a / L^a$

or, in per person terms, $y = Ak^{1/3}$

We can take this model to data, view the data through the lens of the model. We get a = 1/3 by observation of factor shares of some economies. According to Jones (2018, p. 79), from empirical observation, of the US and other countries, a = 1/3 and 1 − a = 2/3, which is a fair approximation. We have data on k and y, but not A.

If A = 1, then the model predicts that predicted y, $y_{PREDICTED} = k^{1/3}$. If predicted y is more than real y, this implies that the parameter A is less than one. Also, the greater the gap between the predicted y and real y, the greater the role of A in overall production, and less of per capita capital.

We now take the model to data and normalize the data with respect to values for the US, which are taken as 1.

We use the `pwt9` package which gives us the Penn World Tables.

```
library(pwt9)
data("pwt9.0")
pwt <- force(pwt9.0)
rm(pwt9.0)
pwt <- as_tibble(pwt)
```

The Penn World Tables has a lot of data relevant to the examination of growth across countries. We focus on the year 2014.

```
pwt2 <- pwt %>%
  filter(year==2014)
```

We select key variables.

```
pwt3 <- pwt2 %>%
  select("cgdpo","emp","pop","ck",
         "country","isocode")
```

We create a variable, output per worker, and another, capital per worker:

```
pwt4 <- pwt3 %>%
  mutate(out_per_worker = cgdpo/pop,
         cap_per_worker = ck/pop )
```

We pull out values for the US:

```
pwt5 <- pwt4 %>%
  filter(isocode=="USA")
USout <- pwt5$out_per_worker
UScap <- pwt5$cap_per_worker
```

We normalize with respect to US values:

```
pwt6 <- pwt4 %>%
  mutate(out = out_per_worker/USout,
         cap = cap_per_worker/UScap,
         pred = cap^(1/3))
```

We see our data for a few select countries:

```
pwt7 <- pwt6 %>%
  filter(country %in% c("United States of America","Japan",
         "Italy", "India", "Brazil", "Spain"))

pwt7 %>%
  select("country","out", "pred", "cap")
## # A tibble: 6 x 4
##    country                    out   pred    cap
##    <fct>                    <dbl>  <dbl>  <dbl>
## 1 Brazil                    0.284  0.737  0.401
## 2 Spain                     0.629  1.03   1.10
## 3 India                     0.104  0.472  0.105
## 4 Italy                     0.676  1.10   1.32
## 5 Japan                     0.683  0.954  0.869
## 6 United States of America 1      1      1
```

We now plot predicted versus real output (Fig. 11.3).

```
ggplot(pwt6, aes(x = out,
                 y = pred)) +
  geom_point(col = "grey40") +
  geom_line(aes(x = out,
                y = out)) +
  scale_x_log10() +
  scale_y_log10()
```

```
## Warning: Removed 2 rows containing missing values
## (geom_point).
```

We observe that there is a large gap between the predicted output and the real output, and it is larger for poorer countries. Poorer countries are poorer not only

Fig. 11.3 Predicted versus
real output

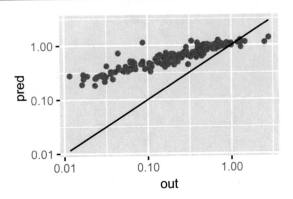

because there is a gap between them and the US because of capital per person, but
also the capital per person is less productive.

11.4   Solow Model Simulation

We will now graph functions of the model and carry out numerical simulations of
the Solow model. The Solow model is usually regarded as a benchmark model by
growth economists. According to Jones (2013, p. 2), 'The modern examination of
this question by macroeconomists dates to the 1950s and the publication of two
famous papers by Robert Solow of the Massachusetts Institute of Technology'.

In the Solow model, we have production given by $Y = AK^aL^{1-a}$ and capital
accumulation given by $K_{t+1} = K_t + sY_t - dK_t$

We generate numbers:

```
a <- 1/3
A <- 2
L <- 200
Klow <- 0
Khigh <- 4000
Knumber <- 100
K <- seq(from = Klow, to = Khigh,
         length.out = Knumber)
Y <- A * (K^a) * (L^(1-a))

Prod <- data.frame(K,Y)
```

We plot Y versus K (Fig. 11.4):

```
ggplot(Prod, aes(x = K,
    y = Y)) +
    geom_line()
```

Fig. 11.4 Y versus K

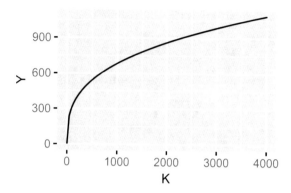

Fig. 11.5 Solow diagram

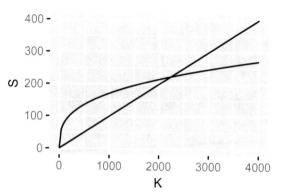

Having plotted the production function, we plot the Solow diagram, which shows how savings and depreciation vary with capital (Fig. 11.5).

```
s <- 0.25
S <- s * Y
d <- 0.1
dep <- d * K

Y <- A * (K^a) * (L^(1-a))

Solow <- data.frame(S,dep,K)
ggplot(Solow) +
  geom_line(aes(x = K, y = S)) +
  geom_line(aes(x = K, y = dep))
```

We can see how if s increases, the Solow diagram and the steady-state values change (Fig. 11.6):

Fig. 11.6 Solow diagram

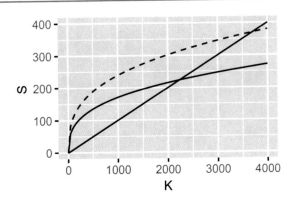

```
s2 <- 0.35
S2 <- s2 * Y
Solow <- data.frame(S,S2, dep, K)
ggplot(Solow) +
  geom_line(aes(x = K, y = S)) +
  geom_line(aes(x = K, y = dep)) +
  geom_line(aes(x = K, y = S2),
           linetype = "dashed")
```

Finally we can plot how the values of Y change over time (Fig. 11.7).

```
d <- 0.1
L <- 200
a <- 1/3
A <- 2
s <- 0.35
d <- 0.1
Kt <- numeric(100)
Kt[1] <- 500
for(i in 2:100) {
  Kt[i] <- Kt[i-1] +
    s * (A * (Kt[i -1]^a) * (L^(1-a)) ) -
    d * Kt[i - 1]
}
Yt <- A * (Kt^a) * (L^(1-a))
motion <- data.frame(Yt,Kt)
ggplot(motion, aes(x = 1:100,
       y = Yt)) +
  geom_line()
```

Fig. 11.7 Y over time

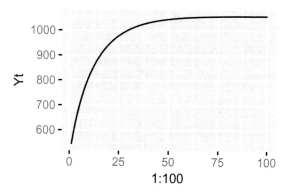

```
ls()
##  [1]  "a"        "A"       "d"        "dep"
##  [5]  "i"        "K"       "Khigh"    "Klow"
##  [9]  "Knumber"  "Kt"      "L"        "mad1"
## [13]  "mad2"     "mad3"    "motion"   "Prod"
## [17]  "pwt"      "pwt2"    "pwt3"     "pwt4"
## [21]  "pwt5"     "pwt6"    "pwt7"     "s"
## [25]  "S"        "s2"      "S2"       "Solow"
## [29]  "UScap"    "USout"   "Y"        "Yt"
rm(list=ls())
ls()
## character(0)
```

11.5 Romer Model Simulation

The Solow model leads to growth that peters out when it depends on capital accumulation, because of diminishing returns.

Romer leads us to ideas as a source of economic growth. Ideas are different from objects. Ideas are characterized by non-rivalry.

Jones (2018, p. 138) clarifies with examples: 'Paul Romer suggested an even more fundamental distinction by dividing the world of economic goods into objects and ideas. Objects include most goods we are familiar with: land, cell phones, oil, jet planes, computers, pencil and paper, as well as capital and labour from the Solow model. Ideas, in contrast, are instructions or recipes. Ideas include designs for making objects. For example, sand (silicon dioxide) has always been of value to beachgoers, kids with shovels, and glass blowers. But with the discovery around 1960 of the recipe for converting sand into computer chips, a new and especially productive use for sand was created. Other ideas include the design of a cell phone or jet engine, the manufacturing technique for turning petroleum into plastic, and the set of instructions for changing trees into paper'.

Fig. 11.8 Y versus Time in
the Romer model

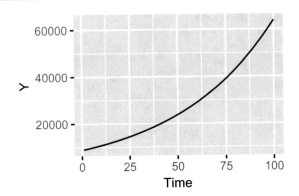

We simulate a simplified version of the Romer model that gives us the essence of
how ideas can lead to growth. Below we use W for ideas and ideas are produced and
are not subject to diminishing returns.

The production function for objects is:

$Y_t = W_t(1 - l)L$, t denoting time and $1 - l$ denoting the share of the total workers
L who work to produce objects.

The production function for ideas is:

$W_t = W_{t-1} + z_t W_{t-1} l L$, where z denotes a productivity parameter and l is the
share of workers who produce ideas. We get continuous growth in the Romer model
(Fig. 11.8).

```
W <- numeric(100)
Y <- numeric(100)
W[1] <- 100
l <- 0.10
L <- 100
z <- 1/500
Y[1] <- W[1] * (1 - l) * L
for(i in 2:100){
  W[i] <- W[i - 1] + z * W[i - 1] * l*L
  Y[i] <- W[i] * (1 - l) * L
}
Romer <- data.frame(W, Y, Time = 1:100)
ggplot(Romer, aes(x = Time,
         y = Y)) +
  geom_line()



rm(list = ls())
```

11.6   Example: Growth in Recent Decades

Here we use the wdi data to examine growth in recent decades. In an earlier chapter we used the WDI package for World Development Indicators data. The initial wrangling follows the same steps.

```
library(WDI)
#new_wdi_cache <- WDIcache()
WDIsearch("gdp.*capita.*PPP")
##        indicator
## [1,] "6.0.GDPpc_constant"
## [2,] "NY.GDP.PCAP.PP.KD.ZG"
## [3,] "NY.GDP.PCAP.PP.KD.87"
## [4,] "NY.GDP.PCAP.PP.KD"
## [5,] "NY.GDP.PCAP.PP.CD"
##        name
## [1,] "GDP per capita, PPP (constant 2011 international $) "
## [2,] "GDP per capita, PPP annual growth (%)"
## [3,] "GDP per capita, PPP (constant 1987 international $)"
## [4,] "GDP per capita, PPP (constant 2011 international $)"
## [5,] "GDP per capita, PPP (current international $)"
#,cache = new_wdi_cache)

wdi_data <- WDI(indicator =
                c("NY.GDP.PCAP.PP.KD"),
                start = 1990,
                end = 2017,
                extra = TRUE)

library(tidyverse)

wdi_data <- wdi_data %>%
  filter(region != "Aggregates")

wdi_data <- wdi_data %>%
  rename(GDP_pc =
           NY.GDP.PCAP.PP.KD)


wdi <- as.tibble(wdi_data)

## Warning: 'as.tibble()' is deprecated, use 'as_tibble()'
(but mind the new semantics).
## This warning is displayed once per session.
```

We now plot boxplots of GDP per capita in 1990, 2004 and 2017 (Fig. 11.9).

Fig. 11.9 Boxplots of GDP per capita (PPP, constant 2011 international $) in 1990, 2004 and 2017

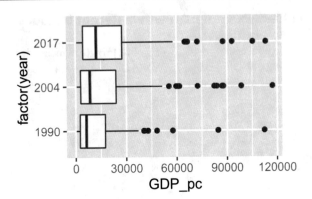

```
wdi_sel <- wdi %>%
  filter(year == 1990 |
           year == 2004 |
           year == 2017)

ggplot(wdi_sel, aes(y = GDP_pc,
                    x = factor(year))) +
  geom_boxplot() +
  coord_flip()

## Warning: Removed 102 rows containing non-finite values
## (stat_boxplot).
```

```
#str(wdi)
```

In Fig. 11.9 we see that the distribution of GDP per capita has shifted to the right. The distribution is positively skewed.

We filter the data for the years 1990 and 2017.

```
wdi_1990 <- wdi %>%
  filter(year == 1990) %>%
  select(country, year, region, GDP_pc,
         iso3c) %>%
  rename(GDP_pc_1990 = GDP_pc)
#str(wdi_1990)

wdi_2017 <- wdi %>%
  filter(year == 2017) %>%
  select(country, year, region, GDP_pc) %>%
  rename(GDP_pc_2017 = GDP_pc)
#str(wdi_2017)
```

We use `left_join()`.

```
wdi_1990_2017 <- wdi_1990 %>%
  left_join(wdi_2017, by = "country")


#str(wdi_1990_2017)
wdi_1990_2017
## # A tibble: 216 x 8
##     country year.x region.x GDP_pc_1990 iso3c
##     <chr>    <int> <fct>          <dbl> <fct>
##  1 Andorra   1990 Europe ~          NA  AND
##  2 United~   1990 Middle ~      112350. ARE
##  3 Afghan~   1990 South A~          NA  AFG
##  4 Antigu~   1990 Latin A~       17473. ATG
##  5 Albania   1990 Europe ~        4458. ALB
##  6 Armenia   1990 Europe ~        3742. ARM
##  7 Angola    1990 Sub-Sah~       4761. AGO
##  8 Argent~   1990 Latin A~      11373. ARG
##  9 Americ~   1990 East As~         NA  ASM
## 10 Austria   1990 Europe ~      31342. AUT
## # ... with 206 more rows, and 3 more variables:
## #   year.y <int>, region.y <fct>,
## #   GDP_pc_2017 <dbl>
```

We now make a scatter plot of Ratio of GDP per capita in 2017 to GDP per capita in 1990 versus GDP per capita in 1990 (Fig. 11.10). For most countries the Ratio of GDP per capita in 2017 to GDP per capita in 1990 lies in the range 1 to 3.

```
library(ggrepel)

wdi_1990_2017 <- wdi_1990_2017 %>%
  mutate(ratio = GDP_pc_2017 / GDP_pc_1990)

ggplot(wdi_1990_2017, aes (x = GDP_pc_1990,
       y = ratio,
       label = iso3c,
       colour = region.x)) +
  geom_text(size = 3) +
  scale_x_log10() +
  scale_y_log10() +
  theme(legend.position = "bottom")


## Warning: Removed 53 rows containing missing values
## (geom_text).
```

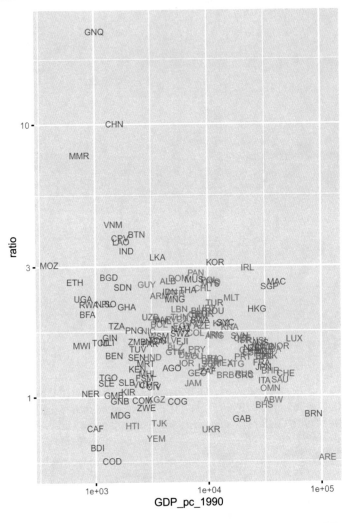

Fig. 11.10 Ratio of GDP per capita in 2017 to GDP per capita in 1990 versus GDP per capita in 1990

We pull out countries with Ratio of GDP per capita in 2017 to GDP per capita in 1990 greater than three (Table 11.1).

```
wdi_1990_2017 %>%
  filter(ratio > 3) %>%
  select(ratio, country) %>%
  arrange(desc(ratio)) %>%
```

Table 11.1 Ratio of GDP per capita in 2017 to GDP per capita in 1990, countries with ratio greater than three

Ratio	Country
21.91	Equatorial Guinea
10.02	China
7.68	Myanmar
4.28	Vietnam
3.93	Bhutan
3.81	Cabo Verde
3.69	Lao PDR
3.42	India
3.24	Sri Lanka
3.09	Korea, Rep.
3.05	Mozambique

Table 11.2 Ratio of GDP per capita in 2017 to GDP per capita in 1990, countries with ratio less than 0.95

Ratio	Country
0.93	The Bahamas
0.91	Zimbabwe
0.86	Madagascar
0.86	Brunei Darussalam
0.83	Gabon
0.80	Tajikistan
0.78	Haiti
0.77	Central African Republic
0.76	Ukraine
0.70	Yemen, Rep.
0.65	Burundi
0.59	United Arab Emirates
0.58	Congo, Dem. Rep.

```
kable(caption = "Ratio of GDP per capita in 2017 to GDP per
    capita in 1990, countries with ratio greater than three",
    digits = 2)
```

We pull out countries with Ratio of GDP per capita in 2017 to GDP per capita in 1990 less than 0.95 (Table 11.2).

```
wdi_1990_2017 %>%
  filter(ratio < 0.95) %>%
  select(ratio, country) %>%
```

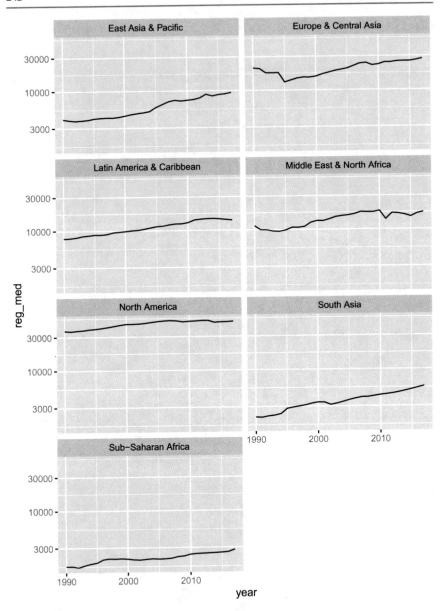

Fig. 11.11 Median GDP per capita by region by year

```
arrange(desc(ratio)) %>%
kable(caption = "Ratio of GDP per capita in 2017 to GDP per
      capita in 1990, countries with ratio less than 0.95",
      digits = 2)
```

We use `group_by` and `summarize` to find the medians of GDP per capita region-wise (Fig. 11.11).

```
wdi_median <- wdi %>%
  group_by(year, region)  %>%
  summarize(reg_med = median(GDP_pc,
                             na.rm = TRUE))
```

We plot the median GDP per capita by region in recent decades.

```
ggplot(wdi_median, aes(x = year,
                       y = reg_med)) +
  geom_line() +
  scale_y_log10() +
  facet_wrap(~ region, ncol = 2)
```

11.7   Resources

Jones (2018) has a clear and accessible presentation of long run growth theory.

Answers to Your Turn

☞ **Your Turn**   Plot the GDP per capita in the USA since 1800, using the filtered tibble mad2. The x variable is year, y variable is gdp_pc. After this plot, plot GDP per capita in the US since 1800 with log scale for the y-axis (scale_y_log10()).

```
ggplot(mad2, aes(x = year, y = gdp_pc)) +
  geom_line()
```

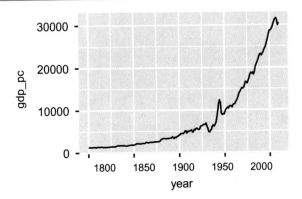

```
ggplot(mad2, aes(x = year, y = gdp_pc)) +
  geom_line() +
  scale_y_log10()
```

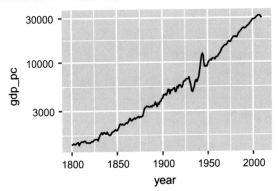

References

Jones, C.I. 2013. Introduction to economic growth. Indian edition. Viva. New Delhi in arrangement with Norton.

Jones, C.I. 2018. *Macroeconomics*. New York: W W Norton and Company.

Persson, E. 2015. Maddison: Maddison project database. *R package version 0.1*. https://CRAN.R-project.org/package=maddison.

Wickham, H. 2017. Tidyverse: Easily install and load the 'Tidyverse'. *R package version 1.2.1*. https://CRAN.R-project.org/package=tidyverse.

Growth Causes

12

12.1 Introduction (Institutions and Growth Example)

Acemoglu, Johnson and Robinson's (AJR, 2004) paper on institutions and growth has been very influential, spawning a mini-literature.

Assenova and Regele (2017) follow up on AJR's paper; they have reproduced AJR's results and also added their own variables. We shall rely on the R code and data available for the paper by Assenova and Regele (2017) in the Harvard dataverse (https://dataverse.harvard.edu).

```
library(tidyverse)
ajr <- read_csv("complete.data.iv.csv")
```

We motivate the issue of the relationship between institutions and growth with a scatter plot of the measure of output versus the measure of institutions, and OLS regressions of output on institutions plus other variables. The measure of output is log GDP per capita PPP in 1995 (`logpgp95`) and the measure of institutions is average risk of expropriation, 1985–1995 (`avexpr`).

We need to filter the data so that we only consider the data used by AJR, with the dummy variable called baseco set to 1:

```
ajrb <- ajr %>%
  filter(baseco == 1)
```

```
library(ggrepel)
ggplot(ajrb, aes(x = avexpr, y = logpgp95,
                 label = shortnam)) +
  geom_text_repel(size = 3) +
  geom_point() +
  geom_smooth(method = "lm", se = FALSE)
```

© Springer Nature Singapore Pte Ltd. 2020
V. Dayal, *Quantitative Economics with R*,
https://doi.org/10.1007/978-981-15-2035-8_12

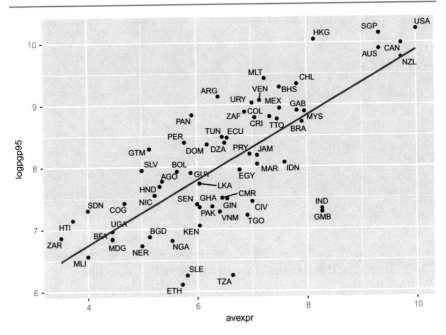

Fig. 12.1 Scatter plot of log GDP per capita PPP in 1995 versus average risk of expropriation, 1985 to 1995

Figure 12.1 shows the scatter plot of log GDP per capita PPP in 1995 versus average risk of expropriation, 1985 to 1995; there is a strong positive relationship.

```
library(texreg)
ol1 <- lm(logpgp95 ~ avexpr, data = ajrb)
ol2 <- lm(logpgp95 ~ avexpr + lat_abst, data = ajrb)
ol3 <- lm(logpgp95 ~ avexpr + lat_abst + asia +
    africa + other, data = ajrb)
texreg(list(ol1, ol2, ol3), caption = "OLS Regressions,
        Dependent variable is log GDP per capita in 1995",
        caption.above = TRUE)
```

Though the correlation is strong as seen in Fig. 12.1 and Table 12.1 (prepared with `texreg`, Leifeld 2013), we expect reverse causality and there are omitted variables. To overcome these issues, AJR propose a theory that can be summarized as follows: *(potential) settler mortality* → *settlements* → *early institutions* → *current institutions* → *current performance.*

They use settler mortality as an instrument for institutions. We plot the reduced form relationship between income and settler mortality (Fig. 12.2); the relationship is strong and negative.

Table 12.1 OLS regressions, dependent variable is log GDP per capita in 1995

	Model 1	Model 2	Model 3
(Intercept)	4.66***	4.73***	5.74***
	(0.41)	(0.40)	(0.40)
avexpr	0.52***	0.47***	0.40***
	(0.06)	(0.06)	(0.06)
lat_abst		1.58*	0.88
		(0.71)	(0.63)
Asia			−0.58*
			(0.23)
Africa			−0.88***
			(0.17)
Other			0.11
			(0.38)
R^2	0.54	0.57	0.71
Adj. R^2	0.53	0.56	0.69
Num. obs.	64	64	64
RMSE	0.71	0.69	0.58

$^*p < 0.05$; $^{**}p < 0.01$; $^{***}p < 0.001$

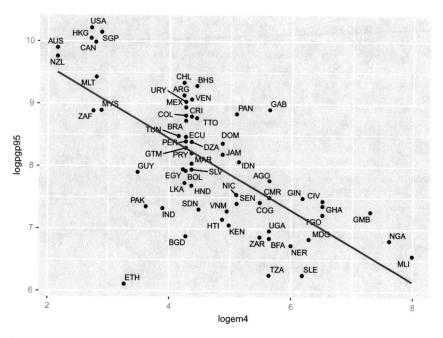

Fig. 12.2 Scatter plot of log GDP per capita PPP in 1995 versus log of mortality rate

```
ggplot(ajrb, aes(x = logem4, y = logpgp95,
                 label = shortnam)) +
  geom_text_repel(size = 3) +
  geom_point() +
  geom_smooth(method = "lm", se = FALSE)
```

We see how average protection against expropriation risk in 1985–1995 is related
to settler mortality (Table 12.2):

```
olA <- lm(avexpr ~ logem4, data = ajrb)
olB <- lm(avexpr ~ logem4 + lat_abst, data = ajrb)
texreg(list(olA, olB), caption = "Dependent variable is
  Average Protection Against Expropriation Risk from 1985 to 1995",
  caption.above = TRUE)
```

We now proceed with the IV estimation (Table 12.3), using the AER package
(Kleiber and Zeileis 2008).

```
library(AER)
mod1 <- ivreg(logpgp95 ~ avexpr |logem4, data =
  ajrb)
mod2 <- ivreg(logpgp95 ~ avexpr +
            lat_abst | logem4 +
        lat_abst, data = ajrb)
mod3 <- ivreg(logpgp95 ~ avexpr + lat_abst + asia +
    africa + other | logem4 + lat_abst + asia +
      africa + other, data = ajrb)
library(texreg)
texreg(list(mod1, mod2, mod3),
  include.rsq = FALSE, include.adjrs = FALSE, include.rmse = FALSE,
  caption = "IV regressions of log GDP per capita",
  caption.above = TRUE)
```

Table 12.2 Dependent variable is average protection against expropriation risk from 1985 to 1995

	Model 1	Model 2
(Intercept)	9.34***	8.53***
	(0.61)	(0.81)
logem4	−0.61***	−0.51***
	(0.13)	(0.14)
lat_abst		2.00
		(1.34)
R^2	0.27	0.30
Adj. R^2	0.26	0.27
Num. obs.	64	64
RMSE	1.26	1.25

$^*p < 0.05; ^{**}p < 0.01; ^{***}p < 0.001$

Table 12.3 IV regressions of log GDP per capita

	Model 1	Model 2	Model 3
(Intercept)	1.91	1.69	1.44
	(1.03)	(1.29)	(2.84)
avexpr	0.94***	1.00***	1.11*
	(0.16)	(0.22)	(0.46)
lat_abst		−0.65	−1.18
		(1.34)	(1.76)
Asia			−1.05
			(0.52)
Africa			−0.44
			(0.42)
Other			−0.99
			(1.00)
Num. obs.	64	64	64

$^*p < 0.05$; $^{**}p < 0.01$; $^{***}p < 0.001$

The effect of average expropriation risk on log GDP per capita is stable across the specifications in Table 12.3 and higher than the OLS estimates in Table 12.1. AJR give the punch line in their paper (p. 1387):

> Let us … compare two 'typical' countries with high and low expropriation risk, Nigeria and Chile (these countries are typical for the IV regression in the sense that they are practically on the regression line). Our 2SLS estimate, 0.94, implies that the 2.24 differences in expropriation risk between these two countries should translate into 206 log point (approximately 7-fold) difference. In practice, the presence of measurement error complicates the interpretation, because some of the difference between Nigeria and Chile's expropriation index may reflect measurement error. Therefore, the 7-fold difference is an upper bound.

12.2 Geography and Growth

Rodrik et al. (2004) followed up on AJR's paper, extending the results. They also considered how geography affects institutions and income levels. They argue that

- geography directly affects income levels and
- geography affects institutions, and institutions and incomes have a bi-directional relationship.

Arguably, the Rodrik et al. (2004) approach is reasonable. However, as they point out, one of the ideas in the literature on economic growth is that geography is the key determinant. The paper by Assenova and Regele (2017) questions AJR. One of the points that they make is that geography is correlated with settler mortality and with economic output (Fig. 12.3). However, we may feel that this is consistent with the ideas of Rodrik et al. (2004).

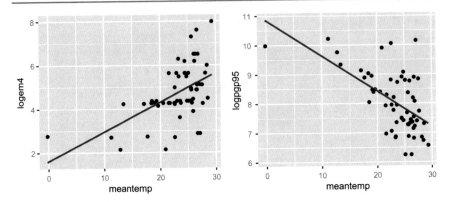

Fig. 12.3 left: Scatter plot of settler mortality versus mean temperature, right: Scatter plot of log GDP per capita versus mean temperature

```
p1 <- ggplot(ajrb, aes(x = meantemp, y = logem4)) +
    geom_point() +
  geom_smooth(method = "lm", se = FALSE)
p2 <- ggplot(ajrb, aes(x = meantemp, y = logpgp95)) +
  geom_point() +
  geom_smooth(method = "lm", se = FALSE)
library(gridExtra)
grid.arrange(p1, p2, ncol = 2)
```

```
## Warning: Removed 4 rows containing non-finite
   values
## (stat_smooth).
## Warning: Removed 4 rows containing missing values
## (geom_point).
## Warning: Removed 4 rows containing non-finite
   values
## (stat_smooth).
## Warning: Removed 4 rows containing missing values
## (geom_point).
```

Geography is also manifest in the continent-wise variation in settler mortality (Fig. 12.4):

```
ggplot(ajrb, aes(x = continent, y = logem4)) +
  geom_boxplot() +
  coord_flip()
```

We can see the continent-wise relationships between mortality and institutions and mortality and income levels below (Figs. 12.5 and 12.6):

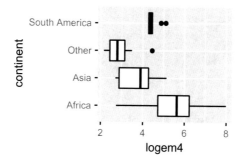

Fig. 12.4 Boxplots of settler mortality by continent

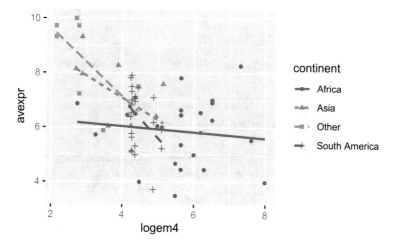

Fig. 12.5 Scatter plot of average risk of expropriation, 1985 to 1995, versus settler mortality

```
ggplot(ajrb, aes(col = continent, x = logem4,
  y = avexpr, shape = continent)) +
  geom_point() +
  geom_smooth(method = "lm", se = FALSE,
              aes(linetype = continent))


ggplot(ajrb, aes(col = continent, x = logem4,
  y = logpgp95, shape = continent)) +
  geom_point() +
  geom_smooth(method = "lm", se = FALSE,
              aes(linetype = continent))
```

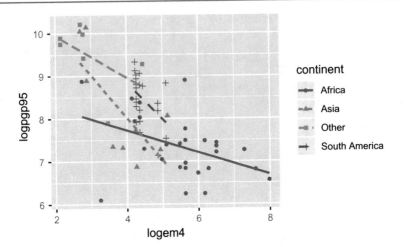

Fig. 12.6 Scatter plot of log GDP per capita PPP in 1995 versus settler mortality

12.3   Exclusion Restriction Simulation

One issue that is raised by Assenova and Regele (2017) is the issue of testing the exclusion restriction of the instrumental variable. Rodrik et al. do not themselves assess the instrumental variable used by AJR, saying that it 'passed what might be called the American Economic Review (AER)-test'.

We cannot directly regress the outcome variable on the endogenous variable and the instrumental variable to test whether the instrument is valid (Morgan and Winship 2014). Even if the instrumental variable is valid, such a regression can show that the coefficient of the instrumental variable is statistically significant, even though we are controlling for the endogenous variable. We demonstrate this with a simulation.

Let U be a common cause of X and Y: $X \leftarrow U \rightarrow Y$; and let Z be a valid instrument of Z, i.e. $Z \rightarrow X \rightarrow Y$.

```
U <- rnorm(300)
Z <- rnorm(300, mean = 3)
X <- Z + 2 * U + rnorm(300)
Y <- X + 2 * U


model1 <- lm(Y ~ X)


library(AER)
model2 <- ivreg(Y ~ X | Z)
texreg(list(model1, model2),
         caption = "OLS of Y on X (Model 1),
IV regression of Y on X (Model 2)",
         caption.above = TRUE)
```

Table 12.4 OLS of Y on X (Model 1), IV regression of Y on X (Model 2)

	Model 1	Model 2
(Intercept)	−2.10***	0.20
	(0.11)	(0.45)
X	1.69***	0.95***
	(0.03)	(0.14)
R^2	0.92	0.75
Adj. R^2	0.92	0.75
Num. obs.	300	300
RMSE	1.12	2.04

$^{*}p < 0.05;\ ^{**}p < 0.01;\ ^{***}p < 0.001$

Table 12.5 Regression of Y on X and Z

	Model 1
(Intercept)	−0.28
	(0.16)
X	1.80***
	(0.02)
Z	−0.72***
	(0.05)
R^2	0.95
Adj. R^2	0.95
Num. obs.	300
RMSE	0.89

$^{*}p < 0.05;\ ^{**}p < 0.01;\ ^{***}p < 0.001$

The regression of Y on X gives us a biased estimate, but the instrumental variable is successful in getting the true effect of X on Y (Table 12.4).

But we find below that though Z is a valid instrumental variable, the regression of Y on X and Z gives us a statistically significant coefficient of Z (Table 12.5):

```
modex3 <- lm(Y ~ X + Z)
texreg(list(modex3),
   caption = "Regression of X on Z",
   caption.above = TRUE)
```

A test that sheds some light on the exclusion assumption is an overidentification test, and this too is done by AJR.

Nevertheless, Deaton (2010) has provided a skeptical but carefully reasoned assessment of some studies using instrumental variables, including that of AJR. He points out that we need to distinguish between a variable that is external (outside the system) and that is exogenous. A variable that is external is necessarily exogenous. Also, regarding the overidentification test, Deaton writes:

> ... because exogeneity is an identifying assumption that must be made prior to analysis of the data, empirical tests cannot settle the question. This does not prevent many attempts in

the literature, often by misinterpreting a satisfactory overidentification test as evidence for valid identification. Such tests can tell us whether estimates change when we select different subsets from a set of possible instruments. While the test is clearly useful and informative, acceptance is consistent with all of the instruments being invalid, while failure is consistent with a subset being correct.

12.4   Other Support for AJR

Acemoglu's study is supported by other observations. In the beginning of their paper, they take the examples of North and South Korea, or East and West Germany as showing the importance of institutions. These cases may be considered to be nature-matched pairs.

In the case of matched pairs created by the analyst using matching, Rosenbaum (2010, p. 322) has suggested that we should take a look at a few matched pairs closely, providing thick descriptions.

Acemoglu and Robinson (2013) begin their book *Why Nations Fail* with a description of the city of Nogales and its two parts: the part in Arizona in the United States, and the part in Sonora, Mexico. The income of the average household in Nogales, Sonora is about one-third that in Nogales, Arizona. There are other differences too, in education, health, etc.

Paul Rosenbaum has written persuasively that though in several instances a study cannot eliminate ambiguity, we can seek to reduce it, especially by drawing on several studies with different strengths and weaknesses. Though ambiguities about the exclusion restriction in the instrumental variable in AJR may remain, other observations and studies with different strengths and weaknesses may convince us that institutions are a key factor in economic growth.

12.5   Resources

Acemoglu and Robinson's (2013) book, *Why Nations Fail* has been praised by several Nobel Laureates in Economics.

Morgan and Winship's (2014) book helps greatly in understanding instrumental variables.

References

Acemoglu, D., and J.A. Robinson. 2013. *Why nations fail*. London: Profile Books.
Acemoglu, D., S. Johnson, and J.A. Robinson. 2001. The colonial origins of comparative development: An empirical investigation. *The American Economic Review* 91 (5): 1369–1401.
Assenova, A.V., and M. Regele. 2017. Revisiting the effect of colonial institutions on comparative economic development. *Plos One* 12 (5): 1–16.
Deaton, A. 2010. Instruments, randomization and learning about development. *Journal of Economic Literature* 48 (June 2010): 424–455.

Kleiber, C., and A. Zeileis. 2008. *Applied econometrics with R*. New York: Springer. https://CRAN. R-project.org/package=AER.

Leifeld, P. 2013. texreg: Conversion of statistical model output in R to LaTeX and HTML tables. *Journal of Statistical Software* 55 (8): 1–24. http://www.jstatsoft.org/v55/i08/.

Morgan, S.L., and C. Winship. 2014. *Counterfactuals and causal inference: Methods and principles for social research.*, Analytical methods for social research Cambridge: Cambridge University Press.

Rodrik, D., A. Subramanian, and F. Trebbi. 2004. Institutions rule: The primacy of institutions over geography and integration in economic development. *Journal of Economic Growth* 9: 131–165.

Rosenbaum, P. 2010. *Design of observational studies.*, Springer series in statistics New York: Springer.

Wickham, H. 2017. Tidyverse: Easily install and load the 'Tidyverse'. *R package version 1.2.1.* https://CRAN.R-project.org/package=tidyverse.

Part VI
Time Series Data

Graphs for Time Series

13

13.1 Introduction

Graphing time series helps us appreciate the different shapes and sizes of time series. How things are changing in the economy can be gauged by examining time series graphs.

13.2 Simple Example with Synthetic Data

The `forecast` package (Hyndman et al. 2019) builds on the `ggplot2` package, so we simply load the `forecast` package. We then create a time series with the `ts` function. The time series starts in 2000 and is annual (frequency $= 1$).

```
library(forecast)
Years5 <- ts(c(138, 91, 54,222,56),
            start = 2000, frequency = 1)
Years5
## Time Series:
## Start = 2000
## End = 2004
## Frequency = 1
## [1] 138  91  54 222  56
```

We use `autoplot()` to plot the time series (Fig. 13.1).

```
autoplot(Years5)
```

© Springer Nature Singapore Pte Ltd. 2020
V. Dayal, *Quantitative Economics with R*,
https://doi.org/10.1007/978-981-15-2035-8_13

Fig. 13.1 Plot of time series

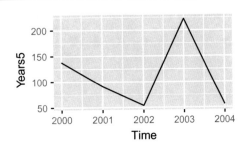

Time

☞ **Your Turn**   We have an annual time series called quanty, which begins in the year 2003, and the values of quanty are: 14, 21, 16, 12, 34. Create a time series object with ts and plot with autoplot.

13.3   Example: Air Passengers

We plot air passenger data described by Cowpertwait and Metcalfe (2009, p. 4), as "The number of international passenger bookings (in thousands) per month on an airline (Pan Am) in the United States ...obtained from the Federal Aviation Administration for the period 1949–1960."

The data is already in R; AirPassengers, we load and rename it.

```
data("AirPassengers")
APass <- AirPassengers
str(APass)
##  Time-Series [1:144] from 1949 to 1961:
      112 118 132 129 121 135 148 148 136 119 ...
class(APass)
## [1] "ts"
head(APass)
##      Jan Feb Mar Apr May Jun
## 1949 112 118 132 129 121 135
autoplot(APass)
```

The Air passenger data show an upward trend with clear seasonality (Fig. 13.2). To have a clearer view of the seasonality, we zoom in on the years after 1958 with the window function and use the ggseasonplot function (Fig. 13.3).

```
ggseasonplot(window(APass, start = 1958))
```

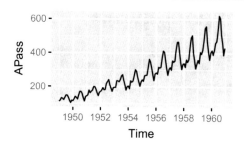

Fig. 13.2 Number of air passengers, thousands per month

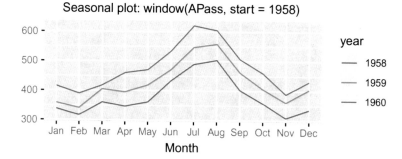

Fig. 13.3 Seasonal plot of air passengers per month, 1958 to 1960

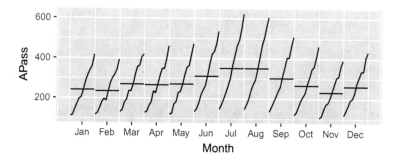

Fig. 13.4 Air passengers per month arranged by month

We can also use the following figure which puts together the data for different years by month:

```
ggsubseriesplot(APass)
```

July and August are the months with the highest number of air passengers, and there is an increase in every month over the years (Fig. 13.4).

13.4   Example: Stock Market Volatility

We use data from Stock and Watson (2011), available in the AER package. Line graphs are excellent tools for describing volatility in the stock market. The time series is the daily New York Stock Exchange stock price index from 1990 to 2005.

```
data("NYSESW", package = "AER")
nyse <- 100 * diff(log(NYSESW))
str(nyse)
## 'zoo' series from 1990-01-03 to 2005-11-11
##    Data: num [1:4002] -0.101 -0.766 -0.844 0.354 -1.019 ...
##    Index:  Date[1:4002], format: "1990-01-03" "1990-01-04"
"1990-01-05" ...
```

We plot the data.

```
autoplot(nyse) +
  geom_smooth(method = "loess")
```

We see that there is considerable volatility, that changes over time (Fig. 13.5). To graph volatility, we estimate the standard deviation at monthly intervals, and use the xts package (Ryan and Ulrich 2018).

```
library(xts)
nyse5 <- as.xts(nyse)
nyse_sd_monthly <- apply.monthly(nyse5, sd)
```

We plot the monthly standard deviations (Fig. 13.6). Volatility was high in the late 1990s and early 2000s.

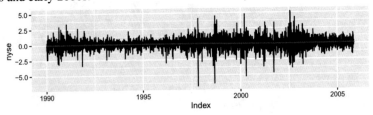

Fig. 13.5 Daily percentage changes in the New York stock exchange index

Fig. 13.6 Monthly standard
deviation of daily percentage
changes in the New York
stock exchange index

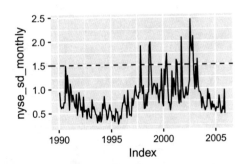

```
autoplot(nyse_sd_monthly) +
  geom_hline(yintercept = 1.5,
             linetype = "dashed")
```

13.5   Example: Inflation and Unemployment

We use data relating to inflation and unemployment used in the book by Stock and
Watson (2011) in the package AER (Kleiber and Zeileis 2008).

```
data("USMacroSW", package = "AER")
```

We can calculate the rate of inflation and bind it to the existing columns,
ts.intersect helps us avoid blank rows since we are using differences. When
we take the difference of the log of the consumer price index (cpi) we get the rate of
inflation. We multiply by 100 to get percentages, and since the data is quarterly we
multiply by 4 to get rates of inflation on a quarterly basis.

```
usm <- ts.intersect
usm <- ts.intersect(USMacroSW,
        4*100*diff(log(USMacroSW[,"cpi"])))
```

We add the name infl (inflation) to the column names.

```
colnames(usm) <- c((colnames(USMacroSW)),"infl")
colnames(usm)
## [1] "unemp"  "cpi"    "ffrate" "tbill"  "tbond"
## [6] "gbpusd" "gdpjp"  "infl"
```

We will now use the tidyverse package (Wickham 2017) for data manipulation. The tsbox package (Sax 2019) helps us convert the time series object to a
time series tibble.

```
library(tidyverse)
library(tsbox)
usm2 <- ts_tbl(usm)
glimpse(usm2)
## Observations: 1,536
## Variables: 3
## $ id    <chr> "unemp", "unemp", "unemp", "une...
## $ time  <date> 1957-04-01, 1957-07-01, 1957-1...
## $ value <dbl> 4.100000, 4.233333, 4.933333, 6...
table(usm2$id)
##
##    cpi ffrate gbpusd  gdpjp   infl  tbill  tbond
##    192    192    192    192    192    192    192
##  unemp
##    192
```

We extract inflation and unemployment, and join them.

```
usm_infl <- usm2 %>%
  filter(id == "infl")
usm_unemp <- usm2 %>%
  filter(id == "unemp")
usm3 <- usm_infl %>%
  left_join(usm_unemp, by = "time")
glimpse(usm3)
## Observations: 192
## Variables: 5
## $ id.x    <chr> "infl", "infl", "infl", "infl...
## $ time    <date> 1957-04-01, 1957-07-01, 1957...
## $ value.x <dbl> 3.3937128, 3.5538940, 1.92951...
## $ id.y    <chr> "unemp", "unemp", "unemp", "u...
## $ value.y <dbl> 4.100000, 4.233333, 4.933333,...
```

We use the `lubridate` package which works with dates and times.

```
library(lubridate)

usm4 <- usm3 %>%
  mutate(Year = year(time) - 1900,
    after_70 = ifelse(Year < 70,
        "70before", "after70"),
        decade = ifelse(Year < 60, "1950s",
          ifelse(Year < 70, "1960s",
            ifelse(Year < 80, "1970s",
              ifelse(Year < 90, "1980s",
                ifelse(Year < 100, "1990s",
                  "2000s")) )))) %>%
  rename(inflation = value.x,
          unemployment = value.y)
```

Is there a tradeoff between inflation and unemployment? Following Leamer (2010), we use graphs to see how the Phillips curve changed over time.

```
ggplot(usm4, aes(x = unemployment,
                 y = inflation)) +
  geom_point() +
geom_smooth(method = "lm", se = FALSE) +
  facet_wrap(~ after_70)
```

We quote Leamer (2010, p. 4), who tells us what the implications of Fig. 13.7 are:

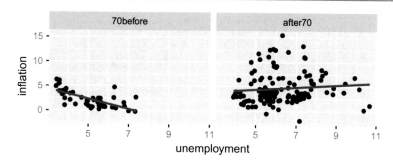

Fig. 13.7 Changing relationship between inflation and unemployment

You can see the Phillips Curve clearly in the figure at the left ... The Phillips curve offered governments some very nasty medicine for unwanted inflation: increase unemployment. The good news for workers around the globe is that the story of the Phillips curve was never very compelling and the subsequent movements in US data made the curve much more difficult to see (see the figure at the right).

We can trace out the path of inflation and unemployment with the geom_path function. In Fig. 13.8 we plot inflation and unemployment and facet by decade. In Fig. 13.9 (top) we plot inflation and unemployment and use colour to differentiate decade. In Fig. 13.9 (bottom) we plot inflation and unemployment points by decade and fit lines, differentiating between decades by colour. The line of fit shifts over decades, and in some periods is flat or positive.

```
ggplot(usm4, aes(x = unemployment,
                 y = inflation,
                 label = Year)) +
  geom_path( arrow =
             arrow(angle = 10,
                   type = "closed",
                   ends = "last")) + facet_wrap(~ decade)

GGi1 <- ggplot(usm4, aes(x = unemployment,
                 y = inflation,
                 label = Year,
                 colour = decade)) +
  geom_path( arrow =
             arrow(angle = 10,
                   type = "closed",
                   ends = "last"))



GGi2 <- ggplot(usm4, aes(x = unemployment,
                 y = inflation,
```

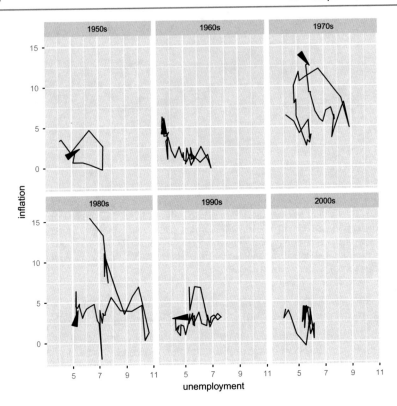

Fig. 13.8 Inflation and unemployment in different decades

```
                      label = Year,
                      colour = decade)) +
        geom_point() +
        geom_smooth(method = "lm", se = FALSE)

library(gridExtra)
grid.arrange(GGi1, GGi2, ncol = 1)
```

13.6   Example: Historical Unemployment Data

Hendry (2015) has compiled macro data related to the United Kingdom dating back
to 1870, and it is available at the book website. We will plot the unemployment data
(Fig. 13.10).

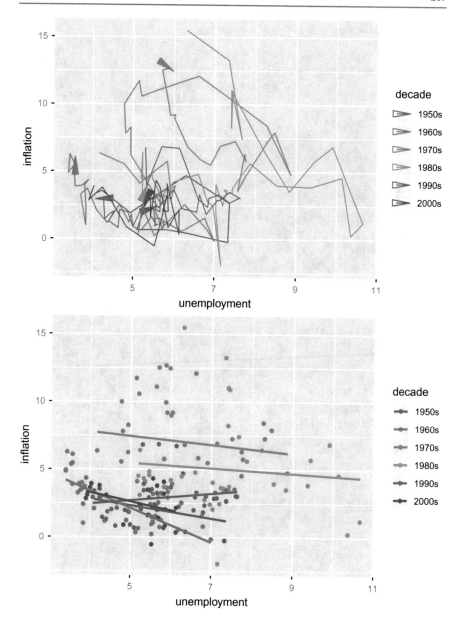

Fig. 13.9 Inflation and unemployment in different decades

Fig. 13.10 Unemployment
in the UK

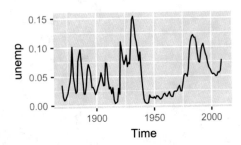

```
MacroData15 <- read_csv("MacroData15.csv")

## Parsed with column specification:
## cols(
##   .default = col_double(),
##   Year = col_character()
## )
## See spec(...) for full column specifications.

unemp <- ts(MacroData15$Ur, start = 1870)
library(forecast)
autoplot(unemp)
```

Time series plots can be very revealing, especially when we annotate the figure.
Below we draw attention to World War I. We use the annotate() function and
build up the plot (Fig. 13.11).

```
p1 <- autoplot(unemp) +
  annotate("text", x = 1916, y = -0.01, label = "World\nWar I") +
    geom_vline(xintercept = c(1914, 1918),
               linetype = "dotted")
p1


p2 <- p1 + annotate("text", x = 1934, y = 0.17,
      label = "Great\nDepression") + ylim(-0.02, 0.18) +
    annotate("text", x = 1942, y = -0.01, label = "World\nWar II") +
    geom_vline(xintercept = c(1939, 1945),
               linetype = "dotted") +
    annotate("text", x = 1973, y = 0.12, label = "Oil\ncrises") +
    annotate("text", x = 1984, y = 0.15, label = "Mrs T")
p2
```

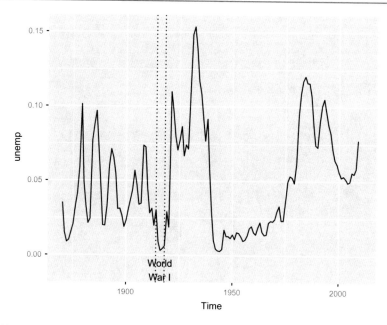

Fig. 13.11 Unemployment in the UK (Graph p1)

☞ **Your Turn** Add in text for "Leave ERM" and "Fin Crisis" using
`annotate()`. Suggested placement for "Leave ERM"
is x = 1992, y = 0.13 and for "Fin Crisis" is x = 2008,
y = 0.11. See the code above as a guide and add to p2 to
create p3 and then print p3.

Hendry (2015) highlights the role of developments outside the economy—in the
two World Wars unemployment was lowest, while economic crises, the oil crisis,
and Mrs. Thatcher's policies were associated with high levels of unemployment
(Fig. 13.12).

13.7 Resources

The forecasting book by Hyndman and Athanasopoulos (2018) has an initial chapter
on plotting time series. Leamer's (2010) book uses a lot of graphs and macroeconomic data; in addition we have insights from someone who has played the role of
econometric skeptic for a long time. Hendry's (2015) book is also graphically rich
while explaining concepts in macro-econometrics, and is available without charge
on the website.

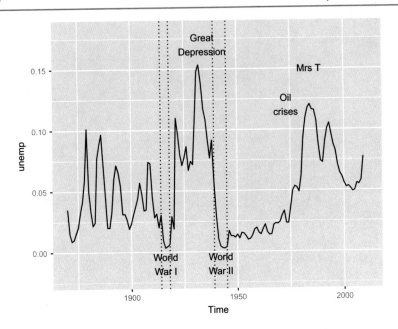

Fig. 13.12 Unemployment in the UK

Your Turn Answers

> ☞ **Your Turn**   Add in text for "Leave ERM" and "Fin Crisis" using
> annotate(). Suggested placement for "Leave ERM"
> is x = 1992, y = 0.13 and for "Fin Crisis" is x = 2008,
> y = 0.11. See the code above as a guide and add to p2 to
> create p3 and then print p3.

```
p3 <- p2 +
  annotate("text", x = 1992, y = 0.13, label = "Leave\nERM") +
  annotate("text", x = 2008, y = 0.11, label = "Fin\nCrisis")
p3
```

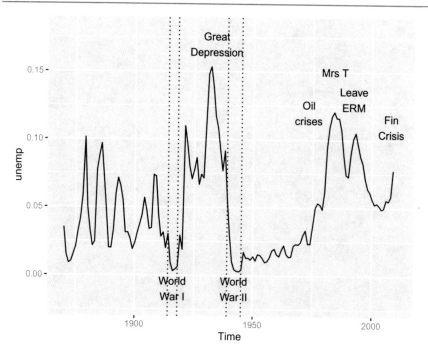

References

Cowpertwait, P.S.P., and A.V. Metcalfe. 2009. Introductory time series with R (Use R!)

Hyndman, R.J., and G. Athanasopoulos. 2018. *Forecasting: principles and practice*, 2nd edition, OTexts: Melbourne. https://OTexts.com/fpp2. Accessed on current date.

Hyndman, R., G. Athanasopoulos, C. Bergmeir, G. Caceres, L. Chhay, M. O'Hara-Wild, F. Petropoulos, S. Razbash, E. Wang, and F. Yasmeen. 2019. forecast: Forecasting functions for time series and linear models. *R Package Version 8.9*. http://pkg.robjhyndman.com/forecast.

Hendry, D. 2015. *Introductory macro-econometrics: A new approach*. London: Timberlake Consultants Ltd. https://www.timberlake.co.uk/intromacroeconometrics. Accessed 28 Oct 2019.

Kleiber, C., and A. Zeileis. 2008. *Applied econometrics with R*. New York: Springer. https://CRAN.R-project.org/package=AER.

Leamer, E. 2010. *Macroeconomic patterns and stories*. New York: Springer.

Ryan, J.A., and J.M. Ulrich. 2018. xts: eXtensible time series. *R Package Version 0.11-2*. https://CRAN.R-project.org/package=xts.

Sax, C. 2019. tsbox: Class-agnostic time series. *R Package Version* (2). https://CRAN.R-project.org/package=tsbox.

Stock, J.H., and M.W. Watson. 2011. *Introduction to econometrics*. Boston: Addison-Wesley.

Wickham, H. 2017. tidyverse: Easily install and load the 'Tidyverse'. *R Package Version* 1 (2): 1. https://CRAN.R-project.org/package=tidyverse.

Basic Time Series Models

14

14.1 Introduction

Time series econometrics is a vast area. Here we deal with basic time series models.

14.2 Simulations

We use simulations to develop intuition about basic building blocks of time series.

14.2.1 White Noise

In a white noise process, the mean and variance are constant, and there is no correlation over time. We generate the white noise process:

```
library(tidyverse)
library(forecast)
white <- rnorm(300)
white_ts <- ts(white, start = 1)
autoplot(white_ts)
```

We have plotted the white noise process using `autoplot()`, `white_ts` in Fig. 14.1; there are no patterns in the data.

We now examine the relationship between the white noise observations and the first lag.

© Springer Nature Singapore Pte Ltd. 2020
V. Dayal, *Quantitative Economics with R*,
https://doi.org/10.1007/978-981-15-2035-8_14

Fig. 14.1 White noise
process

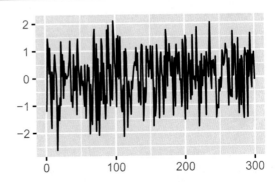

Fig. 14.2 Scatter plot of
white noise versus first lag of
white noise

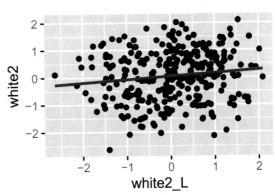

```
white2 <- white[-300]
white2_L <- white[-1]
white_lag <- tibble(white2, white2_L)
glimpse(white_lag)
## Observations: 299
## Variables: 2
## $ white2   <dbl> -1.20771130, 0.02040129, 1.4...
## $ white2_L <dbl> 0.02040129, 1.45935819, 1.12...
```

We make a scatter plot of white noise versus first lag of white noise (Fig. 14.2).

```
ggplot(white_lag, aes(x = white2_L,
                      y = white2)) +
  geom_point() +
  geom_smooth(method = "lm", se = FALSE)
```

```
cor(white2_L, white2)
## [1] 0.1174224
```

Figure 14.2 shows that there is no relationship between the white noise observations and first lag of white noise. The lack of correlation between present and

Fig. 14.3 ACF of white
noise

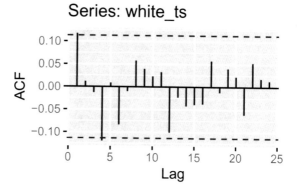

lagged values of `white_ts` is reflected in the plot (Fig. 14.3) of the autocorrelation
function (ACF):

```
ggAcf(white_ts)
```

Another way of simulating white noise is with the `arima.sim` function (arima
is autoregressive integrated moving average) (Fig. 14.4):

```
autoplot(arima.sim(300, model = list(ar = 0, ma = 0)))
```

We can estimate the white noise model with the `Arima` function:

```
mod_w <- Arima(white_ts, order=c(0,0,0))
mod_w
## Series: white_ts
## ARIMA(0,0,0) with non-zero mean
##
## Coefficients:
##            mean
##          0.0556
## s.e.     0.0533
##
## sigma^2 estimated as 0.8556:  log likelihood=-401.78
## AIC=807.56    AICc=807.6    BIC=814.97


mean(white_ts)
## [1] 0.05561418
var(white_ts)
## [1] 0.8555541
```

Fig. 14.4 White noise simulated with arima.sim()

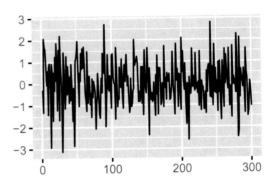

Fig. 14.5 Autoregressive process

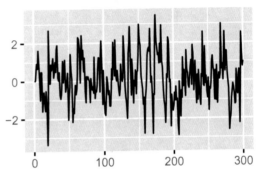

Fig. 14.6 Autoregressive process

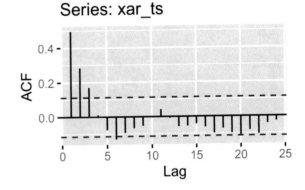

☞ **Your Turn**   Simulate 300 observations of a white noise process with mean = 3 and sd = 2, using the `arima.sim` function. Then, use the `Arima` function to estimate the white noise model.

Fig. 14.7 checkresiduals()

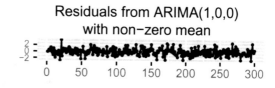

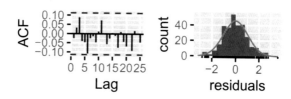

Fig. 14.8 Random walk

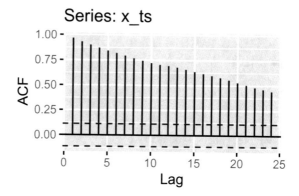

Fig. 14.9 ACF for random walk

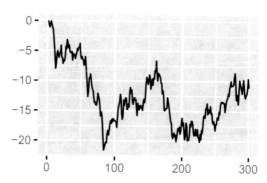

14.2.2 Autoregressive Model

A simple example of an autoregressive process is:
$Today = 0.5 * yesterday + white\_noise$
We generate the data.

Fig. 14.10 Differenced
random walk

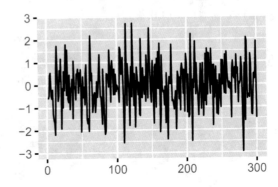

Fig. 14.11 ACF of
differenced random walk

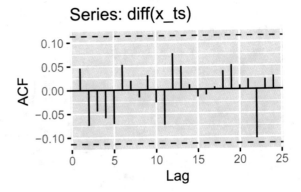

```
# generate the data
xar <- numeric(300)
w <- rnorm(300)
for (t in 2:300) {
xar[t] <- 0.5 * xar[t -1] + w[t]
}
xar_ts <- ts(xar, start = 1)
```

We plot (Figs. 14.5 and 14.6).

```
autoplot(xar_ts)
```

```
ggAcf(xar_ts)
```

The Acf shows a high value at lag 1 consistent with the data generation process.
We use Arima() to fit a model.

```
mod1 <- Arima(xar_ts, order = c(1, 0, 0))
mod1
## Series: xar_ts
## ARIMA(1,0,0) with non-zero mean
##
## Coefficients:
##           ar1    mean
##        0.4943  0.1007
## s.e.   0.0500  0.1187
##
## sigma^2 estimated as 1.096:  log likelihood=-438.6
## AIC=883.2    AICc=883.28    BIC=894.31
```

We use the checkresiduals() function to check the residuals. If the residuals resemble white noise, our fit is good (Fig. 14.7).

```
checkresiduals(mod1)

##
##  Ljung-Box test
##
## data:  Residuals from ARIMA(1,0,0) with non-zero mean
## Q* = 8.4866, df = 8, p-value = 0.3874
##
## Model df: 2.   Total lags used: 10
```

The Ljung–Box test p-value is just above 0.05 so we do not reject the hypothesis that the residuals are independently distributed.

We can also generate an autoregressive process using the arima.sim function.

☞ **Your Turn**   Run the following code (remove the #). What do you observe?

```
#library(forecast)
#xar2 <- arima.sim(list(ar = -0.9), n = 300)
#autoplot(xar2)
#ggAcf(xar2)
#ggPacf(xar2)
```

14.2.3 Random Walk

A random walk model is an example of a non-stationary process. A random walk can be expressed simply as:

$Today = yesterday + noise$

We generate the data, and plot (Figs. 14.8 and 14.9).

```
# generating random walk
x <- numeric(300)
w <- rnorm(300)
for (t in 2:300) {
  x[t] <- x[t-1] + w[t]
}
# converting to ts
x_ts <- ts(x)
autoplot(x_ts)
```

```
ggAcf(x_ts)
```

The Acf decays slowly.

We get white noise if we difference a random walk (Figs. 14.10 and 14.11).

```
autoplot(diff(x_ts))
```

```
ggAcf(diff(x_ts))
```

We generate another two random walks and then plot three random walks together (Fig. 14.12); this helps us appreciate the nature of random walks, as suggested by the name.

```
x2 <- numeric(300)
w <- rnorm(300)
for (t in 2:300) {
  x2[t] <- x2[t-1] + w[t]
}
```

```
x3 <- numeric(300)
w <- rnorm(300)
for (t in 2:300) {
  x3[t] <- x3[t-1] + w[t]
}
```

```
RW_ts <- ts(cbind(x3,x2,x))
autoplot(RW_ts)
```

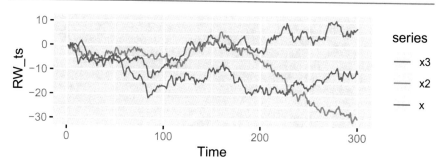

Fig. 14.12 Three random walks

14.2.4 Moving Average

We will simulate a moving average model (Figs. 14.13 and 14.14):
$$Today = noise + 0.3 * yesterday's\ noise$$

```
# generate data
xma <- numeric(300)
xma[1] <- 0
w <- rnorm(300)
for (t in 2:300) {
  xma[t] <- w[t] + 0.3 * w[t - 1]
}

xma_ts <- ts(xma)
autoplot(xma_ts)


ggAcf(xma_ts)
```

We can fit a model:

```
summary(Arima(xma_ts, order = c(0,0,1)))
## Series: xma_ts
## ARIMA(0,0,1) with non-zero mean
##
## Coefficients:
##          ma1       mean
##       0.3116   -0.0246
## s.e.  0.0571    0.0737
##
## sigma^2 estimated as 0.9545:  log likelihood=-417.75
## AIC=841.5   AICc=841.58   BIC=852.61
##
## Training set error measures:
```

```
##                         ME         RMSE        MAE MPE
## Training set 0.0003430139 0.9737367 0.7808742 Inf
##                      MAPE       MASE         ACF1
## Training set     Inf 0.8084612 -0.005053989
```

14.2.5 Autoregressive Moving Average

An autoregressive moving average process (arma) combines autoregressive and moving average components. We simulate an arma process (Fig. 14.15):

```
arma1 <- arima.sim(n = 200, list(order = c(1,0,1), ar = c(0.6) ,
        ma = c(0.4))) + 20
autoplot(arma1)
```

We plot the ACF (Fig. 14.16).

```
ggAcf(arma1)
```

We fit a model, and check residuals (Fig. 14.17).

```
modarma <- Arima(arma1, order = c(1,0,1))
modarma
## Series: arma1
## ARIMA(1,0,1) with non-zero mean
##
## Coefficients:
##            ar1       ma1      mean
##         0.5143    0.4434   20.2857
## s.e.    0.0759    0.0776    0.2181
##
## sigma^2 estimated as 1.109:  log likelihood=-293.05
## AIC=594.1   AICc=594.3   BIC=607.29
checkresiduals(modarma)
```

```
##
##   Ljung-Box test
##
## data:  Residuals from ARIMA(1,0,1) with non-zero mean
## Q* = 8.0463, df = 7, p-value = 0.3285
##
## Model df: 3.   Total lags used: 10
```

We also use the auto.arima function, which uses the Hyndman–Khandakar algorithm (Hyndman and Athanasopoulos 2018) to fit an arima model of suitable order (Fig. 14.18).

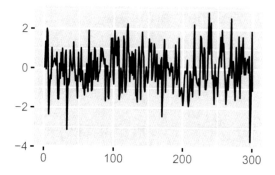

Fig. 14.13 Moving average process

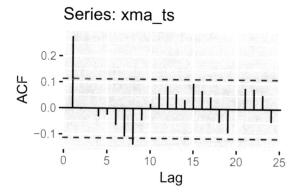

Fig. 14.14 ACF of moving average model

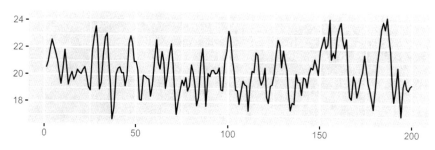

Fig. 14.15 Plot of arma process

```
modarma.auto <- auto.arima(arma1)
modarma.auto
## Series: arma1
## ARIMA(1,0,1) with non-zero mean
##
## Coefficients:
##             ar1       ma1       mean
```

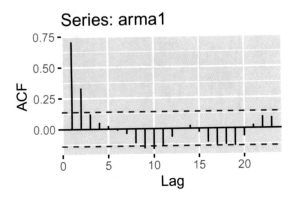

Fig. 14.16 ACF of arma process

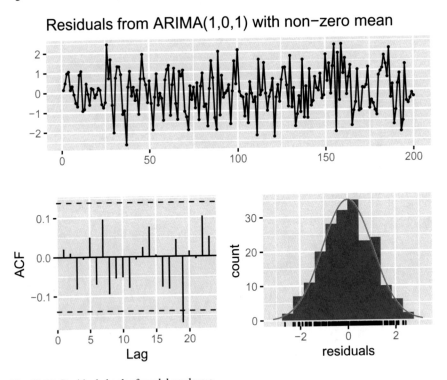

Fig. 14.17 Residual check of model modarma

```
##         0.5143  0.4434   20.2857
## s.e.    0.0759  0.0776    0.2181
##
## sigma^2 estimated as 1.109:  log likelihood=-293.05
## AIC=594.1   AICc=594.3   BIC=607.29
checkresiduals(modarma.auto)
```

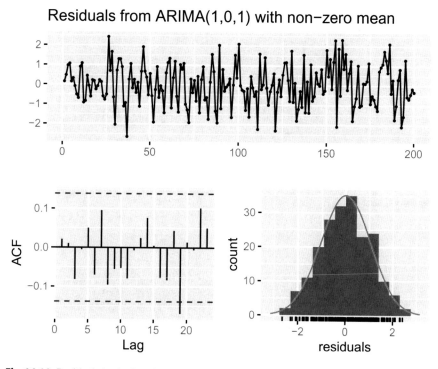

Fig. 14.18 Residual check of model modarma.auto

```
##
##   Ljung-Box test
##
## data:   Residuals from ARIMA(1,0,1) with non-zero mean
## Q* = 8.0463, df = 7, p-value = 0.3285
##
## Model df: 3.    Total lags used: 10
```

14.3   Example: Forecasting Inflation

We will now forecast inflation using the auto.arima function in the forecast package. This function incorporates an algorithm (see Hyndman and Athanasopoulos 2018).

```
data("USMacroSW", package = "AER")

usm <- ts.intersect
usm <- ts.intersect(USMacroSW,
                    4*100*diff(log(USMacroSW[,"cpi"])))
```

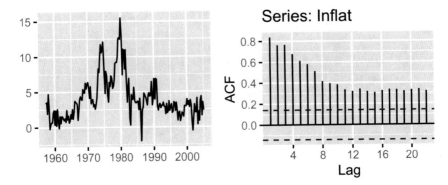

Fig. 14.19 Inflation

```
colnames(usm) <- c((colnames(USMacroSW)),"infl")
colnames(usm)
## [1] "unemp"  "cpi"     "ffrate" "tbill"  "tbond"
## [6] "gbpusd" "gdpjp"  "infl"
Inflat <- usm[,"infl"]


str(Inflat)
##  Time-Series [1:192] from 1957 to 2005: 3.39 3.55 1.93 4.71
##               2.68 ...
```

We plot inflation and its acf (Fig. 14.19):

```
library(gridExtra)
gr1 <- autoplot(Inflat)
gr2 <- ggAcf(Inflat)
grid.arrange(gr1, gr2, ncol = 2)
```

The acf indicates non-stationarity as it decays very slowly. We can use a formal test. In the Dickey–Fuller test, we test whether $\beta_1 = 1$ in the equation $Y_t = \beta_0 + \beta_1 Y_{t-1} + u_t$.

The Augmented Dickey–Fuller test is better—it augments the Dickey–Fuller test by lags of the difference of Y.

We install and load the package tseries; then use the function *adf.test*.

```
#install.packages("tseries")
library(tseries)
adf.test(Inflat)
##
##  Augmented Dickey-Fuller Test
##
## data:  Inflat
## Dickey-Fuller = -2.5724, Lag order = 5,
```

```
## p-value = 0.3366
## alternative hypothesis: stationary
```

Our null hypothesis that a unit root is present was not rejected, so we difference inflation to achieve stationarity, and check (Figs.14.20 and 14.21).

```
Infl_diff <- diff(Inflat)
autoplot(Infl_diff)
```

```
ggAcf(Infl_diff)
```

```
adf.test(Infl_diff)
```

```
## Warning in adf.test(Infl_diff): p-value smaller than
printed p-value
```

```
##
##   Augmented Dickey-Fuller Test
##
## data:  Infl_diff
## Dickey-Fuller = -6.1122, Lag order = 5,
## p-value = 0.01
## alternative hypothesis: stationary
```

The null hypothesis of a unit root is now rejected.

A simple way to forecast a series Z is to use its past values. For example, if we estimate a first-order autoregressive model,

$$Z_t = \beta_0 + \beta_1 Z_{t-1} + u_t$$

we can get a forecast for Z for a period ahead with

$$Z_{T+1} = \beta_0^{EST} + \beta_1^{EST} Z_T,$$

where EST denotes estimate.

We might find that our forecasts improve with more lags than one.

We use the auto.arima function, which is an algorithm that selects a suitable arima model.

```
mod <- auto.arima(Inflat,
    D = 0, max.Q = 0, max.P = 0)
mod
## Series: Inflat
## ARIMA(3,1,0)
##
## Coefficients:
##              ar1       ar2       ar3
```

Fig. 14.20 Difference of inflation

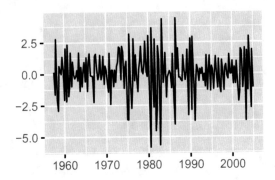

Fig. 14.21 Difference of inflation

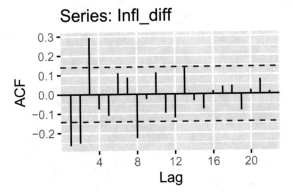

```
##          -0.3033  -0.2959   0.1464
## s.e.      0.0717   0.0715   0.0723
##
## sigma^2 estimated as 2.35:  log likelihood=-351.29
## AIC=710.57    AICc=710.79   BIC=723.58
```

We check the residuals (Fig. 14.22):

```
checkresiduals(mod)
```

```
##
##   Ljung-Box test
##
## data:  Residuals from ARIMA(3,1,0)
## Q* = 9.4698, df = 5, p-value = 0.09173
##
## Model df: 3.    Total lags used: 8
```

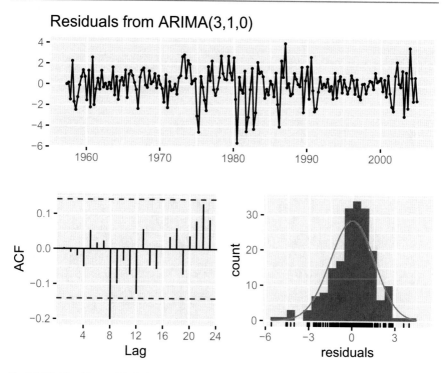

Fig. 14.22 Checking residuals from mod

We get the forecast values:

```
forecast(mod, level = 95)
##             Point Forecast      Lo 95     Hi 95
## 2005 Q2         1.757215  -1.247155  4.761586
## 2005 Q3         2.554563  -1.107080  6.216205
## 2005 Q4         2.326057  -1.623607  6.275721
## 2006 Q1         2.070284  -2.538960  6.679529
## 2006 Q2         2.332236  -2.754728  7.419201
## 2006 Q3         2.295001  -3.123223  7.713226
## 2006 Q4         2.191332  -3.637788  8.020452
## 2007 Q1         2.272153  -3.929014  8.473320
```

We plot the forecast (Fig. 14.23):

```
mod %>% forecast(h = 6) %>%
   autoplot()
```

We need to use our judgment about which past values we will use for forecasting. According to Shmueli and Lichtendahl Jr (2016, p. 41), 'While a very short (and recent) series might be insufficiently informative for forecasting purposes, beyond a certain length the additional information will likely be useless at best, and harmful

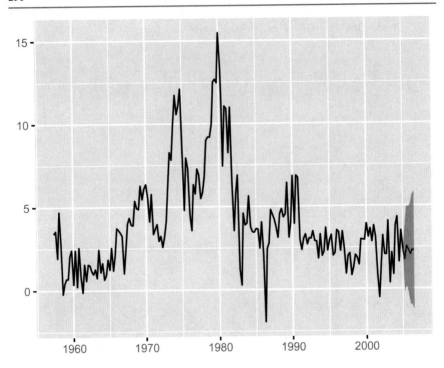

Fig. 14.23 Forecast of inflation (mod)

at worst. Considering a very long past of the series might deteriorate the accuracy of future forecasts because of the changing context and environment occurring during the data period'. If the environment has changed, we may choose to use only a portion of the historical data. For example we could, in this example, use values from 1992 we then check the residuals (Fig. 14.24).

```
mod2 <- auto.arima(window(Inflat, start = 1992), D = 0,
    max.Q = 0, max.P = 0)
mod2
## Series: window(Inflat, start = 1992)
## ARIMA(0,0,0) with non-zero mean
##
## Coefficients:
##          mean
##        2.5135
## s.e.   0.1394
##
## sigma^2 estimated as 1.05:   log likelihood=-75.99
## AIC=155.98    AICc=156.22    BIC=159.92
checkresiduals(mod2)
```

Fig. 14.24 mod2 for inflation

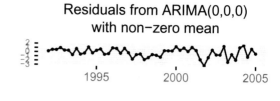

Residuals from ARIMA(0,0,0)
with non−zero mean

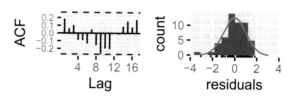

```
##
##   Ljung-Box test
##
## data:  Residuals from ARIMA(0,0,0) with non-zero mean
## Q* = 6.5954, df = 7, p-value = 0.4722
##
## Model df: 1.   Total lags used: 8
forecast(mod2, level = 95)
##           Point Forecast      Lo 95      Hi 95
## 2005 Q2        2.513537  0.5052596  4.521814
## 2005 Q3        2.513537  0.5052596  4.521814
## 2005 Q4        2.513537  0.5052596  4.521814
## 2006 Q1        2.513537  0.5052596  4.521814
## 2006 Q2        2.513537  0.5052596  4.521814
## 2006 Q3        2.513537  0.5052596  4.521814
## 2006 Q4        2.513537  0.5052596  4.521814
## 2007 Q1        2.513537  0.5052596  4.521814
```

We plot the forecast (Fig. 14.25):

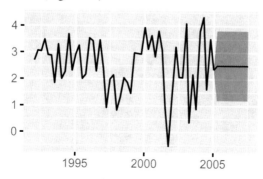

Fig. 14.25 Forecast of inflation (mod 2)

```
mod2 %>% forecast(h = 10) %>%
   autoplot()
```

14.4   Cointegration

14.4.1 Simulating Spurious Regression

Regressing two non-stationary variables that are actually not related can give us spurious results. We generate two random walks, x.col and y.col.

```
set.seed(29)
x.col <- rnorm(100)
y.col <- rnorm(100)
for (i in 2:100) {
   x.col[i] <- x.col[i - 1] + rnorm(1)
   y.col[i] <- y.col[i - 1] + rnorm(1)
}
```

Though y.col and x.col are not related the estimated coefficient of x.col in the regression of y.col on x.col is statistically significant.

We see our spurious regression, with high t-values for the coefficient:

```
mod.col <- lm(y.col ~ x.col)
summary(mod.col)
##
## Call:
## lm(formula = y.col ~ x.col)
##
## Residuals:
##     Min      1Q Median      3Q     Max
## -5.344  -2.238  -1.075   2.471   7.140
##
## Coefficients:
##               Estimate Std. Error t value Pr(>|t|)
## (Intercept)    1.53196    0.54535   2.809    0.006
## x.col         -0.24496    0.04391  -5.579 2.15e-07
##
```

Fig. 14.26 Spurious relationship example

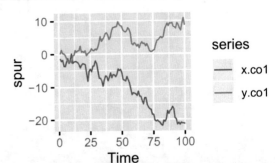

Fig. 14.27 Spurious
relationship example

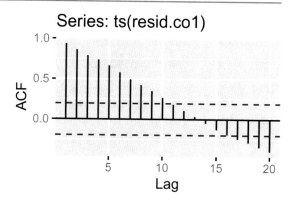

Series: ts(resid.co1)

```
## (Intercept) **
## x.co1        ***
## ---
## Signif. codes:
## 0 '***' 0.001 '**' 0.01 '*' 0.05 '.' 0.1 ' ' 1
##
## Residual standard error: 3.016 on 98 degrees of freedom
## Multiple R-squared:  0.2411, Adjusted R-squared:  0.2333
## F-statistic: 31.13 on 1 and 98 DF,  p-value: 2.145e-07
resid.co1 <- mod.co1$resid
```

The separate stochastic trends in x.co1 and y.co1 lead to spurious regression. The two variables appear related (Fig. 14.26).

```
spur <- ts(cbind(x.co1,y.co1))

autoplot(spur)

ggAcf(ts(resid.co1))
```

The residual Acf graph shows persistence (Fig. 14.27).

14.4.2 Simulating Cointegration

We generate two variables that have a common stochastic trend.

```
set.seed(46)

x.co2 <- y.co2 <- rw <- numeric(300)
rw[1] <- 2
for (i in 2:300) {
  rw[i] <- rw[i - 1] + rnorm(1)
  x.co2[i] <- rw[i -1] + rnorm(1)
```

Fig. 14.28 ACF of residuals
from cointegrated variables

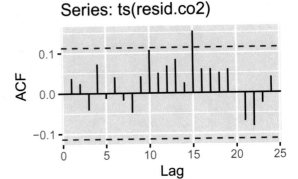

Fig. 14.29 Federal funds
rate

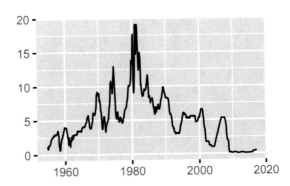

```
    y.co2[i] <- rw[i -1] + rnorm(1)
}

co2.ts <- ts(x.co2,y.co2)
```

When we regress the two variables, we get residuals with minimal autocorrelation
(Fig. 14.28).

```
mod.co2 <- lm(y.co2 ~ x.co2)
summary(mod.co2)
##
## Call:
## lm(formula = y.co2 ~ x.co2)
##
## Residuals:
##     Min      1Q  Median      3Q     Max
## -4.3764 -0.8647 -0.0433  0.9774  5.4602
##
## Coefficients:
##              Estimate Std. Error t value Pr(>|t|)
## (Intercept)  0.03413    0.11140   0.306     0.76
## x.co2        0.97980    0.02170  45.147   <2e-16
##
```

```
## (Intercept)
## x.co2          ***
## ---
## Signif. codes:
## 0 '***' 0.001 '**' 0.01 '*' 0.05 '.' 0.1 ' ' 1
##
## Residual standard error: 1.548 on 298 degrees of freedom
## Multiple R-squared:  0.8724, Adjusted R-squared:  0.872
## F-statistic:  2038 on 1 and 298 DF,  p-value: < 2.2e-16
resid.co2 <- mod.co2$resid
ggAcf(ts(resid.co2))
```

We can test for cointegration formally with po.test(); for x.co2 and y.co2 the null hypothesis of no cointegration is rejected.

```
library(tseries)
co2.b <- as.matrix(cbind(x.co2, y.co2))
po.test(co2.b)
```

```
## Warning in po.test(co2.b): p-value smaller than
printed p-value
```

```
##
##  Phillips-Ouliaris Cointegration Test
##
## data:  co2.b
## Phillips-Ouliaris demeaned = -286.09,
## Truncation lag parameter = 2, p-value = 0.01
```

For x.co1 and y.co1 the null hypothesis of no cointegration is not rejected.

```
co1.b <- as.matrix(cbind(x.co1, y.co1))
po.test(co1.b)
```

```
## Warning in po.test(co1.b): p-value greater than
printed p-value
```

```
##
##  Phillips-Ouliaris Cointegration Test
##
## data:  co1.b
## Phillips-Ouliaris demeaned = -3.2333,
## Truncation lag parameter = 0, p-value = 0.15
```

Fig. 14.30 Bond rate

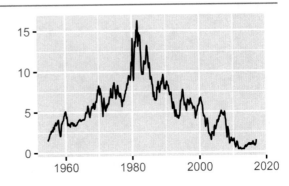

14.4.3 Example: Federal Funds and Bond Rate

We use an example from Hill et al. (2018), presented in Colonescu (2018). We consider two variables here: (1) the federal funds rate (the interest rate on overnight loans between banks), `ffr`, and (2) the three-year bond rate (interest rate on a financial asset to be held for three years), `br`.

```
##devtools::install_github("ccolonescu/POE5Rdata")
library(POE5Rdata) #activates the package "POE5Rdata"
usa.ts <- ts(usdata5, start = c(1954, 8),
             end = c(2016, 12),
             frequency = 12)
```

We plot the two series.

```
autoplot(usa.ts[,"ffr"])
```

```
autoplot(usa.ts[,"br"])
```

The two series do appear to move together (Figs. 14.29, 14.30). We test for cointegration:

```
cobind <- as.matrix(cbind(usa.ts[,"ffr"], usa.ts[,"br"]))
po.test(cobind)

## Warning in po.test(cobind): p-value smaller than
printed p-value

##
##   Phillips-Ouliaris Cointegration Test
##
## data:  cobind
## Phillips-Ouliaris demeaned = -65.376,
## Truncation lag parameter = 7, p-value = 0.01
```

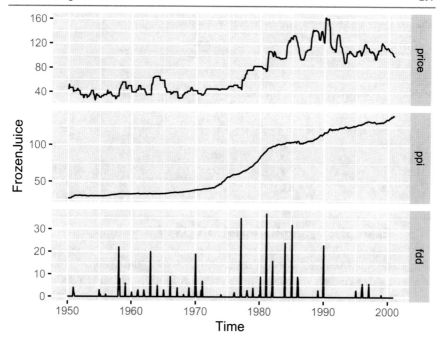

Fig. 14.31 Series in Frozen Juice data

The null hypothesis of no cointegration is rejected.

```
ffr_ts <- usa.ts[,"ffr"]
br_ts <- usa.ts[, "br"]
modc <- lm(br_ts ~ ffr_ts)
```

We regress the bond rate on the federal funds rate (Table 14.1).

```
library(texreg)
texreg(list(modc),
       caption = "Regression of bond rate on the federal funds
                  rate.",
       caption.above = TRUE)
```

Hill et al. (p. 583) observe:

'The result–that the federal funds and bond rates are cointegrated–has major economic implications! It means that when the Federal Reserve implements monetary policy by changing
the federal funds rate, the bond rate will also change thereby ensuring that the effects of
monetary policy are transmitted to the rest of the economy'.

Table 14.1 Regression of bond rate on the federal funds rate

	Model 1
(Intercept)	1.33***
	(0.06)
ffr_ts	0.83***
	(0.01)
R^2	0.91
Adj. R^2	0.91
Num. obs.	749
RMSE	0.96

$^{*}p < 0.05,\ ^{**}p < 0.01,\ ^{***}p < 0.001$

14.5   Example: Dynamic Causal Effects of Weather

In this example from Stock and Watson (2011), we see the *dynamic causal effect* of freezing degree days on orange juice prices in Florida. Freezing weather harms orange trees. However, orange juice can be stored, so the effect of freezing weather plays out over time, so we investigate the dynamic causal effect. Stock and Watson provide a clear intuitive discussion of estimation of dynamic causal effects. They argue that freezing degree days are exogenous, (p. 591) 'From the perspective of orange juice markets, we can think of the weather–the number of freezing degree days–as if it were randomly assigned, in the sense that the weather is outside human control. If the effect of FDD is linear and if it has no effect on prices after r months, then it follows that the weather is exogenous'.

```
library("dynlm")
library("AER")
data("FrozenJuice", package = "AER")
head(FrozenJuice)
##            price     ppi fdd
## Jan 1950  43.6 27.20000   0
## Feb 1950  52.1 27.20000   0
## Mar 1950  46.6 27.20000   0
## Apr 1950  46.6 27.20000   0
## May 1950  46.6 27.29924   0
## Jun 1950  46.6 27.39962   0
```

The dataset contains monthly data on the price of frozen orange juice concentrate (price), the producer price index (ppi) and the number of freezing degree days (fdd) for Florida. The freezing degree days measure captures both whether the temperature went below freezing and by how much (see Fig. 14.31). The temperature went below freezing and by how much (see Fig. 14.31).

```
autoplot(FrozenJuice, facet = TRUE)
```

We do an initial regression of the percentage change in prices in a month on the number of freezing degree days in that month.

```
fm_dyn <- dynlm(d(100 * log(price/ppi)) ~ fdd,
    data = FrozenJuice)
coeftest(fm_dyn, vcov = vcovHC(fm_dyn),
        type = "HC1")
##
## t test of coefficients:
##
##               Estimate Std. Error t value Pr(>|t|)
## (Intercept) -0.42095    0.18954 -2.2209 0.026723
## fdd          0.46724    0.15006  3.1137 0.001934
##
## (Intercept) *
## fdd          **
## ---
## Signif. codes:
## 0 '***' 0.001 '**' 0.01 '*' 0.05 '.' 0.1 ' ' 1
```

One more freezing degree day during a month increases the price of orange juice concentrate in that month by 0.47%.

Next, we run a distributed lag regression (Table 14.2).

```
fm_dyn2 <- dynlm(d(100 * log(price/ppi)) ~
                L(fdd, 0:6),
            data = FrozenJuice)
pack <- coeftest(fm_dyn2, vcov = vcovHC(fm_dyn2),
        type = "HC1")
library(broom)
kable(tidy(pack), digits = 2,
    caption = "Effects of frozen degree days on orange juice
            prices")
```

An extra freezing degree day is estimated to increase the orange juice concentrate price in the next month by 0.15%, two months later by 0.06% and three months later by 0.07% (Table 14.2).

14.6   Resources

For Better Understanding

The following Datacamp courses (datacamp.com) are excellent:

- Introduction to Time Series Analysis by David Matteson,
- ARIMA Modeling with R by David Stoffer and
- Forecasting Using R by Rob J Hyndman.

Table 14.2 Effects of frozen degree days on orange juice prices

Term	Estimate	Std. error	Statistic	p-value
(Intercept)	−0.69	0.21	−3.30	0.00
L(fdd, 0:6)0	0.47	0.15	3.13	0.00
L(fdd, 0:6)1	0.15	0.09	1.59	0.11
L(fdd, 0:6)2	0.06	0.07	0.89	0.37
L(fdd, 0:6)3	0.07	0.05	1.45	0.15
L(fdd, 0:6)4	0.04	0.03	1.07	0.29
L(fdd, 0:6)5	0.05	0.03	1.41	0.16
L(fdd, 0:6)6	0.05	0.05	0.95	0.34

Fig. 14.32 White noise simulated with arima.sim

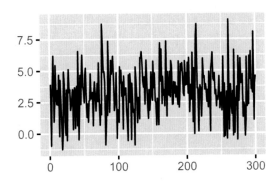

Stock and Watson (2011) is a fine econometrics text written by leading time series econometricians, with very good chapters on time series. Hill et al. (2018) is also approachable and has good chapters on time series econometrics. Cowpertwait and Metcalfe (2009) is a book that introduces time series with R; highly recommended.

To Go Further

Hyndman and Athanasopoulos (2018) is a detailed but accessible text on diverse approaches to forecasting that uses R. Hendry and Nielsen's (2007) text on econometric modelling devotes about half its pages to time series econometrics, and you can feel the energy of Hendry's exposition coming through.

Answers to Your Turn

☞ **Your Turn**   Simulate 300 observations of a white noise process with mean = 3 and sd = 2, using the `arima.sim` function. Then, use the `Arima` function to estimate the white noise model.

```
white_yt <- arima.sim(300, model = list(ar =0, ma = 0), mean = 3,
          sd = 2)
autoplot(white_yt)


mod_wyt <- Arima(white_yt, order=c(0,0,0))
mod_wyt
## Series: white_yt
## ARIMA(0,0,0) with non-zero mean
##
## Coefficients:
##          mean
##        3.3007
## s.e.   0.1107
##
## sigma^2 estimated as 3.691:  log likelihood=-621.08
## AIC=1246.17   AICc=1246.21   BIC=1253.58
```

See Fig. 14.32.

References

Colonescu, C. 2018. *Using R for principles of econometrics*. Scotts Valley: Createspace Independent Publisher.

Cowpertwait, P.S.P., A.V. Metcalfe. 2009. *Introductory Time Series with R*. London: Springer.

Hendry, D., and B. Nielsen. 2007. *Econometric Modeling: A Likelihood Approach*. Princeton: Princeton University Press.

Hill, R.C., W.E. Griffiths, and G.C. Lim. 2018. *Principles of Econometrics*. Hoboken: Wiley.

Hyndman, R.J., G. Athanasopoulos. 2018. *Forecasting: principles and practice*, 2nd edition, OTexts: Melbourne. http://OTexts.com/fpp2. Accessed on 27 2019.

Hyndman, R., G. Athanasopoulos, C. Bergmeir, G. Caceres, L. Chhay, M. O'Hara-Wild, F. Petropoulos, S. Razbash, E. Wang, F. Yasmeen. 2019. forecast: Forecasting functions for time series and linear models. R package version 8.9. http://pkg.robjhyndman.com/forecast.

Shmueli, G., K.C. Lichtendahl, Jr. 2016. *Practical time series forecasting with r: a hands-on guide*. Green Cove Springs: Axelrod Schnall Publishers.

Stock, J.H., M.W. Watson. 2011. *Introduction to Econometrics*. Boston: Addison-Wesley.

Part VII

Introduction to Statistical/Machine Learning from Data

Smoothers and Generalized Additive Models

15

15.1 Introduction

A key substantive assumption that we make when we use a linear model is linearity. Smoothers and generalized additive models help us relax this assumption.

15.2 Simple Example with Synthetic Data

We generate data, where y is a nonlinear function of x.

```
library(tidyverse)
x <- seq(from = 0, to = 11, length.out = 100)
y <- -x * (10 - x) + 3 * rnorm(100)
gam1 <- tibble(x, y)
```

We plot y against x.

```
ggplot(gam1, (aes(x = x, y = y))) +
  geom_point() +
  geom_smooth(method = "lm")
```

There is a lack of fit (Fig. 15.1). We can use a polynomial regression fit, because of the obvious quadratic nature of the curve here (Fig. 15.2):

```
ggplot(gam1, (aes(x = x, y = y))) +
  geom_point() +
  geom_smooth(method = "lm", formula =
                y ~ poly(x,2))
```

© Springer Nature Singapore Pte Ltd. 2020
V. Dayal, *Quantitative Economics with R*,
https://doi.org/10.1007/978-981-15-2035-8_15

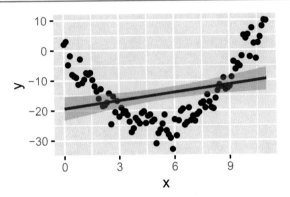

Fig. 15.1 Scatter plot of y against x, with linear fit

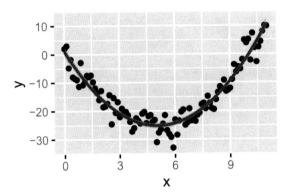

Fig. 15.2 Scatter plot of y against x, with polynomial fit

We can plot this again, with a loess smoother (Fig. 15.3):

```
ggplot(gam1, (aes(x = x, y = y))) +
  geom_point() +
  geom_smooth(method = "loess")
```

Loess is a nonparametric smoother. Loess, local regression, computes the line of fit at a target point, say x_0 using only points in the neighbourhood of x_0, which are weighted, with points further from x_0 being weighted less. To get some broad intuition about loess, we fit separate lines to regions of x (Fig. 15.4).

```
ggplot(gam1, (aes(x = x, y = y,
         shape = cut(x, 5)))) +
  geom_point() +
  geom_smooth(method = "lm")
```

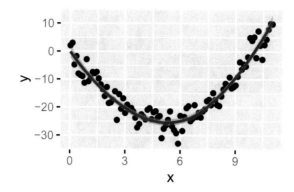

Fig. 15.3 Scatter plot of y against x, with loess fit

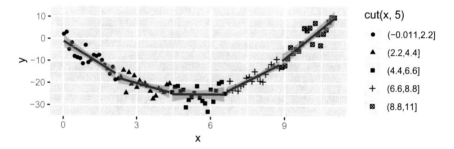

Fig. 15.4 Scatter plot of y against x, with piece-wise linear fit

15.3   Example: GAMS with Wages Data

GAMs are generalized additive models. A GAM allows us to extend the multiple linear regression model. If we have a regression model:

$$y_i = \beta_0 + \beta_1 x_{i1} + \beta_2 x_{i2} + \epsilon_i,$$

we can write the GAM as:

$$y_i = \beta_0 + f_1(x_{i1}) + f_2(x_{i2}) + \epsilon_i.$$

We replace each linear component $\beta_j x_{ij}$ with a smooth non-linear function $f_j(x_{ij})$ (James et al. 2013).

We use an example from James et al. (2013). The Wage data set has data on wages and other variables from the Atlantic region of the United States.

```
library(ISLR)
data("Wage", package = "ISLR")
str(Wage)
## 'data.frame':    3000 obs. of  11 variables:
##  $ year       : int  2006 2004 2003 2003 2005 2008 2009 2008
##                      2006 2004 ...
##  $ age        : int  18 24 45 43 50 54 44 30 41 52 ...
##  $ maritl     : Factor w/ 5 levels "1. Never Married",..: 1 1
##                      2 2 4 2 2 1 1 2 ...
```

```
##  $ race      : Factor w/ 4 levels "1. White","2. Black",..: 1
##                  1 1 3 1 1 4 3 2 1 ...
##  $ education : Factor w/ 5 levels "1. < HS Grad",..: 1 4 3 4
##                  2 4 3 3 3 2 ...
##  $ region    : Factor w/ 9 levels "1. New England",..: 2 2 2
##                  2 2 2 2 2 2 ...
##  $ jobclass  : Factor w/ 2 levels "1. Industrial",..: 1 2 1 2
##                  2 2 1 2 2 2 ...
##  $ health    : Factor w/ 2 levels "1. <=Good","2. >=Very Good":
##                  1 2 1 2 1 2 2 1 2 2 ...
##  $ health_ins: Factor w/ 2 levels "1. Yes","2. No": 2 2 1 1 1
##                  1 1 1 1 1 ...
##  $ logwage   : num  4.32 4.26 4.88 5.04 4.32 ...
##  $ wage      : num  75 70.5 131 154.7 75 ...
```

We first visualize the data:

```
ggplot(Wage, aes(x = age, y = wage)) +
  geom_point(alpha = 0.5, col = "grey70")  +
  geom_smooth()
```

```
## 'geom_smooth()' using method = 'gam' and formula 'y
~ s(x, bs = "cs")'
```

Wages first curve up and then down with age (Fig. 15.5).

```
ggplot(Wage, aes(y = wage, x = factor(education))) +
  geom_boxplot()  +
  coord_flip()
```

Higher education is associated with greater wages (Fig. 15.6).

```
ggplot(Wage, aes(y = wage, x = factor(year))) +
  geom_boxplot()  +
  coord_flip()
```

Fig. 15.5 Scatter plot of wage versus age

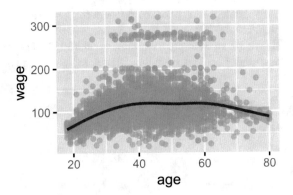

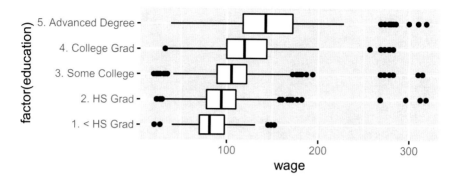

Fig. 15.6 Boxplots of wage by education category

Fig. 15.7 Boxplots of wage by year

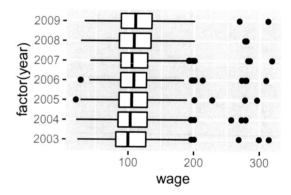

Wages increase over time (Fig. 15.7).

We now fit a generalized additive model with gam(), using the gam package (Hastie 2019).

```
library(gam)
Wage$fac_year <- factor(Wage$year)
Wage$fac_education <- factor(Wage$education)
gamMod <- gam(wage ~ lo(age) + fac_year +
                    fac_education, data = Wage)

plot(gamMod, se = TRUE, pch =".")
```

In Fig. 15.8 we see the that Wage increases with age and then levels off a bit and then falls. Figure 15.9 shows the increase of Wage with year, and Fig. 15.10 shows the increase of Wage with education.

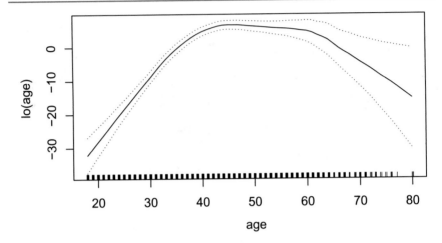

Fig. 15.8 gamMod

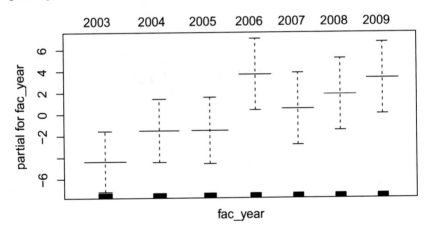

Fig. 15.9 gamMod

15.4   Example: Housing in Texas

We now see how loess can help us see annual and seasonal trends in data. We use data from the ggplot2 package (Wickham 2016). We have monthly data from January 2000 to April 2015, on the number of house sales (sales) for 46 Texas cities.

```
library(forecast)
library(ggplot2)
data("txhousing")
texas <- txhousing
str(texas)
## Classes 'tbl_df', 'tbl' and 'data.frame':     8602 obs. of
##     9 variables:
## $ city      : chr  "Abilene" "Abilene" "Abilene" "Abilene" ...
```

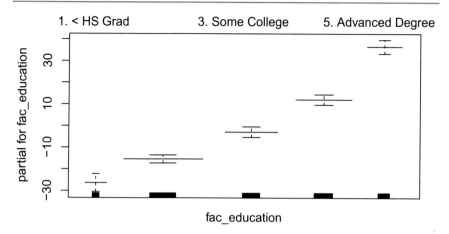

Fig. 15.10 gamMod

```
##   $ year     : int   2000 2000 2000 2000 2000 2000 2000 2000
##                      2000 2000 ...
##   $ month    : int   1 2 3 4 5 6 7 8 9 10 ...
##   $ sales    : num   72 98 130 98 141 156 152 131 104 101 ...
##   $ volume   : num   5380000 6505000 9285000 9730000
##                      10590000 ...
##   $ median   : num   71400 58700 58100 68600 67300 66900 73500
##                      75000 64500 59300 ...
##   $ listings : num   701 746 784 785 794 780 742 765 771 764 ...
##   $ inventory: num   6.3 6.6 6.8 6.9 6.8 6.6 6.2 6.4 6.5 6.6 ...
##   $ date     : num   2000 2000 2000 2000 2000 ...
```

We focus on one city, Abilene, we filter the data. We plot sales versus date, with a loess smooth (Fig. 15.11).

```
rm(txhousing)
library(tidyverse)
tex_abil <- texas %>%
  filter(city == "Abilene")
ggplot(tex_abil, aes(date, sales)) +
  geom_line() +
  geom_smooth(method = loess)
```

There is an upward trend in sales, along with seasonality. We convert the sales data for Abilene to a time series.

```
tex_sales_ts <- ts(tex_abil$sales,
                   frequency = 12,
                   start = 2000)
```

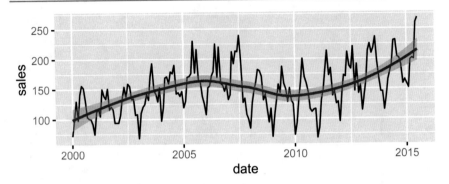

Fig. 15.11 Line plot of sales versus data with loess smooth

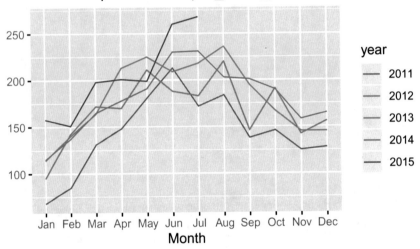

Fig. 15.12 Season plot of sales

To get a closer look at the seasonal data, we use the `ggseasonplot()` function. Sales appear to be lowest in January, and rise to high levels in May–August (Fig. 15.12).

```
ggseasonplot(window(tex_sales_ts,
                    start = 2011))
```

We use the Seasonal and Trend decomposition using Loess (STL) method available in the `forecast` package (Hyndman et al. 2019). The time series data is decomposed into trend, seasonal and remainder components (Fig. 15.13).

```
autoplot(stl(tex_sales_ts, s.window = 7))
```

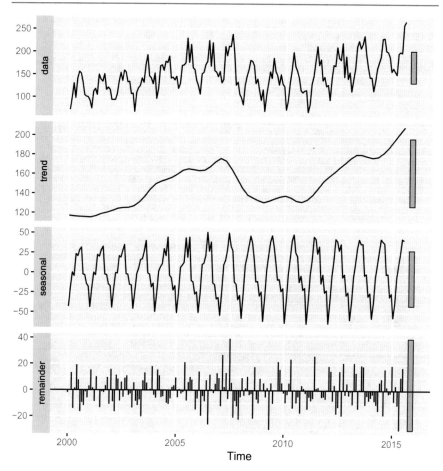

Fig. 15.13 Time series decomposition of Abilene housing sales

15.5 Resources

James et al. (2013) is clear and accessible, with labs and a corresponding online course by Hastie and Tibshirani.

References

Hastie, T. 2019. gam: generalized additive models. *R package version* 1 (16): 1. https://CRAN.R-project.org/package=gam.

Hyndman, R., G. Athanasopoulos, C. Bergmeir, G. Caceres, L. Chhay, M. O'Hara-Wild, F. Petropoulos, S. Razbash, E. Wang, F. Yasmeen. 2019. forecast: Forecasting functions for time series and linear models. *R package version 8.9*. http://pkg.robjhyndman.com/forecast.

James, G., D. Witten, T. Hastie, and R. Tibshirani. 2013. *An introduction to statistical learning: with applications in R.* New York: Springer.

Wickham, H. 2016. *ggplot2: elegant graphics for data analysis*, 2016. New York: Springer.

From Trees to Random Forests

16

16.1 Introduction

Varian (2014) points out the increasing availability of a large amount of data on economic transactions because of computers. He writes (p. 3), 'Machine learning techniques …may allow for more effective ways to model complex relationships. …my standard advice to graduate students these days is to go to the computer science department and take a class in machine learning'.

Breiman (2001, p. 199) writes: 'There are two cultures in the use of statistical modeling to reach conclusions from data. One assumes that the data are generated by a given stochastic data model. The other uses algorithmic models and treats the data mechanism as unknown. …If our goal as a field is to use data to solve problems, then we need to move away from exclusive dependence on data models and adopt a more diverse set of tools'.

Trees are a nonparametric computationally intensive technique for prediction, where the predictor space is recursively split into regions. Trees are easily interpretable, but can be unstable. One way to improve prediction is to combine trees. We first get a feel for trees, and then consider random forests.

Biau and Scornet (p. 197) endorse our attention to random forests: 'The random forest algorithm, proposed by L. Breiman in 2001, has been extremely successful as a general-purpose classification and regression method. The approach, which combines several randomized decision trees and aggregates their predictions by averaging, has shown excellent performance in settings where the number of variables is much larger than the number of observations. Moreover, it is versatile enough to be applied to large-scale problems, is easily adapted to various ad hoc learning tasks, and returns measures of variable importance'.

© Springer Nature Singapore Pte Ltd. 2020
V. Dayal, *Quantitative Economics with R*,
https://doi.org/10.1007/978-981-15-2035-8_16

16.2   Simple Tree Example with Synthetic Data

We create some synthetic data and then plot to see its features.

```
x.tree <- c(rep(1:5,20), rep(6:10,20),
       rep(11:15,20))
y.tree <- c(rep(c(0,1,0,0,1),20),
       rep(c(1,0,1,1,1),20),
       rep(c(0,0,0,0,1),20))
xy.tree <- data.frame(x.tree, y.tree)

library(tidyverse)

# plotting the data
ggplot(xy.tree, aes(x = x.tree, y = y.tree)) +
  geom_jitter(height = 0.1, width = 0.1) +
  geom_smooth()
```

The relationship between y.tree and x.tree is very nonlinear (Fig. 16.1). We will now fit a classification tree to the data. We use the rpart package (Therneau and Atkinson 2019). We use the same syntax that we use for regression.

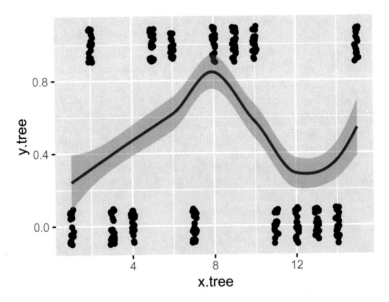

Fig. 16.1 Scatter plot of y.tree versus x.tree

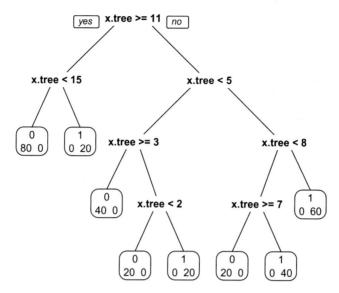

Fig. 16.2 Classification tree for dataset xy.tree

```
library(rpart)
xy.tree_t <- rpart(y.tree ~ x.tree,
            data = xy.tree,
            method = "class")
class(xy.tree_t)
## [1] "rpart"
```

We now use the `rpart.plot` package (Milborrow 2019) to plot the tree that we fitted.

```
library(rpart.plot)
prp(xy.tree_t, extra = 1)
```

Figure 16.2 is the plot of the classification tree. It predicts that y.tree will be one or zero depending on the value of x.tree; we go down the tree according to the value of x.tree. If x.tree is 9, for example, we start at the top, x.tree is not greater than or equal to 11, so we move down the tree to the right. At the next step, x.tree ($= 9$) is not less than 5 so again we move down to the right. x.tree ($= 9$) is not less than 8, so we move down to the right. The prediction is that y.tree for x.tree $= 9$ is 1. Had x.tree been 8 or 10 we would have followed the same path down the tree. There are 60 observations with x.tree $= 8, 9$ or 10, and all have a value of y.tree of 1.

We now look at an example concerning arsenic in wells in Bangladesh.

16.3   Example: Arsenic in Wells in Bangladesh

We present an example where we have a binary response, and economists would usually use logistic regression in this situation. We will fit a logistic regression, then a tree to the same data, so that we can connect the tree to a more familiar method.

Gelman and Hill (2017) present an example of arsenic in wells in Bangladesh. In this region, some wells have high levels of arsenic, which is a cumulative poison. A research team had measured the level of arsenic and labelled the wells; then they returned a few years later to see who had switched to a safer well by conducting a survey.

We will use the example for illustration, and so we focus on the following variables: (1) `switch`: if the household switched or not, (2) `dist`: the `distance` in metres to the closest known safe well and (3) `arsenic`: the `arsenic` level of the respondent's well (Table 16.1).

```
wells <- read.delim("~/Documents/R/ies2018/wells.dat",
                  header = TRUE, sep = "")
library(tidyverse)
```

We fit a logistic regressions of switch on distance and arsenic.

```
fit <- glm(switch ~ dist + arsenic,
   family = binomial( link = logit),
   data = wells)
library(texreg)
texreg(list(fit), caption = "Logistic regression of switch
        on distance and arsenic",
        caption.above = TRUE)
```

Table 16.1 Logistic regression of switch on distance and arsenic

	Model 1
(Intercept)	0.00
	(0.08)
dist	−0.01***
	(0.00)
arsenic	0.46***
	(0.04)
AIC	3936.67
BIC	3954.71
Log likelihood	−1965.33
Deviance	3930.67
Num. obs.	3020

$^*p < 0.05, ^{**}p < 0.01, ^{***}p < 0.001$

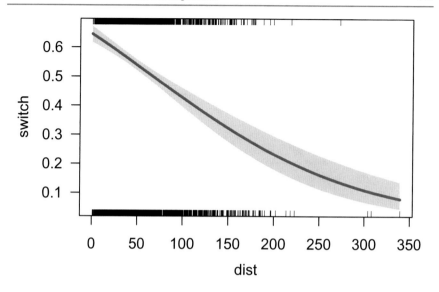

Fig. 16.3 Predicted probability of switching versus distance

The visreg package (Breheny and Burchett 2017) is especially useful for plotting nonlinear regressions.

```
library(visreg)
visreg(fit,"dist",
       scale = "response")


visreg(fit,"arsenic",
       scale = "response")
```

The logistic regressions (Figs. 16.3 and 16.4) are on the lines of what is expected. In Fig. 16.3, the predicted probability of switching falls from about 0.6, when distance is close to zero, to about 0.1 when distance is about 275. In Fig. 16.4, the probability of switching increases from about 0.5 when arsenic is close to zero to about 0.95 when arsenic is close to 8.

We now fit a classification tree.

```
library(rpart)
Bang <- rpart(switch ~ dist +
              arsenic,
         data = wells,
         method = "class")

library(rpart.plot)
prp(Bang, extra = 1)
```

In Fig. 16.5 we see the classification tree. At the top we have a split at arsenic less than 1.1. If no, and if distance is less than 82, then we predict that switching would occur. If arsenic is not less than 1.1 and distance is greater than or equal to 82, switching will not occur.

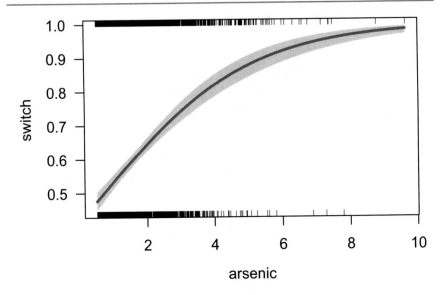

Fig. 16.4 Predicted probability of switching versus arsenic

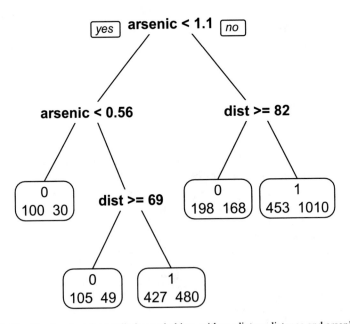

Fig. 16.5 Classification tree for predicting switching, with predictors distance and arsenic

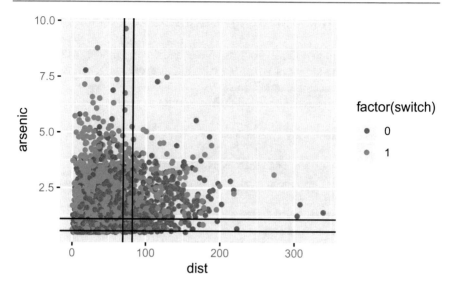

Fig. 16.6 Scatter plot of arsenic and distance, with regions corresponding to classification tree in Fig. 16.5

If arsenic is less than 0.56, then the prediction is of not switching.

A tree like that in Fig. 16.5 is easily interpretable.

We use a scatter plot of arsenic versus distance and plot the different regions in the predictor space corresponding to the classification tree splits (Fig. 16.6).

```
ggplot(wells, aes(x = dist,
                  y = arsenic,
                  colour = factor(switch))) +
  geom_point() +
  geom_hline(yintercept = 1.1) +
  geom_hline(yintercept = 0.56) +
  geom_vline(xintercept = 69) +
  geom_vline(xintercept = 82)
```

☞ **Your Turn**   Look at Figs. 16.5 and 16.6 closely, and see how they relate to each other. Try to relate the splits in the trees in Fig. 16.5 to different rectangular regions in Fig. 16.6.

16.4   Example: Home Mortgage Disclosure Act

In this section we will first fit a classification tree and then apply the random forest method. Although the output of a tree is easily interpretable, in general, a random forest will give more accurate predictions.

Varian (2014, p.14) has summarized the random forest method:

Random forests is a technique that uses multiple trees. A typical procedure uses the following steps.

1. Choose a bootstrap sample of the observations and start to grow a tree.

2. At each node of the tree, choose a random sample of the predictors to make the next decision. Do not prune the trees.

3. Repeat this process many times to grow a forest of trees.

4. In order to determine the classification of a new observation, have each tree make a classification and use a majority vote for the final prediction.

We now see how we can predict whether mortgage applicants in Boston would be denied (variable deny) mortgage lending or not, on the basis of several predictors. The data is available in the Ecdat package.

```
library(Ecdat)
data(Hdma)
names(Hdma)[11] <- "condo"
set.seed(111)
```

```
library(tidyverse)
hdma <- Hdma %>%
  na.omit()
glimpse(hdma)
## Observations: 2,380
## Variables: 13
## $ dir    <dbl> 0.221, 0.265, 0.372, 0.320, 0....
## $ hir    <dbl> 0.221, 0.265, 0.248, 0.250, 0....
## $ lvr    <dbl> 0.8000000, 0.9218750, 0.920398...
## $ ccs    <dbl> 5, 2, 1, 1, 1, 1, 1, 2, 2, 2, ...
## $ mcs    <dbl> 2, 2, 2, 2, 1, 1, 2, 2, 2, 1, ...
## $ pbcr   <fct> no, no, no, no, no, no, no, no...
## $ dmi    <fct> no, no, no, no, no, no, no, no...
## $ self   <fct> no, no, no, no, no, no, no, no...
## $ single <fct> no, yes, no, no, no, no, yes, ...
## $ uria   <dbl> 3.9, 3.2, 3.2, 4.3, 3.2, 3.9, ...
## $ condo  <dbl> 0, 0, 0, 0, 0, 0, 1, 0, 0, 0, ...
## $ black  <fct> no, no, no, no, no, no, no, no...
## $ deny   <fct> no, no, no, no, no, no, no, no...
```

We fit a classification tree.

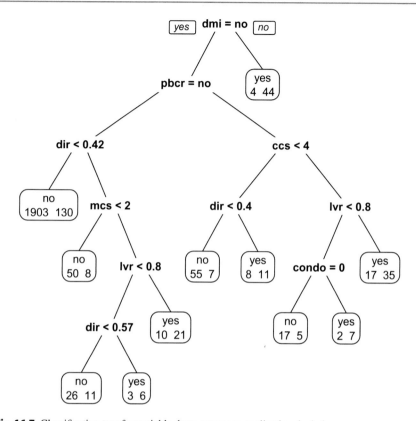

Fig. 16.7 Classification tree for variable deny, mortgage application denied

```
library(rpart)
hm.tree <- rpart(deny ~., data = hdma,
                 method = "class")
```

We plot the classification tree (Fig. 16.7).

```
library(rpart.plot)
prp(hm.tree, extra = 1)
```

The top split in the tree is of the variable dmi—denied mortgage insurance. The next variable in the tree is pbcr (public bad credit record), followed by dir (debt payments to total income ratio) and ccs (consumer credit score).

We now use the random forest algorithm.

```
library(randomForest)
set.seed(1234)
rf.fit <- randomForest(deny ~ ., data = hdma,    importance = TRUE)
```

```
rf.fit
##
## Call:
##   randomForest(formula = deny ~ ., data = hdma, importance = TRUE)
##                  Type of random forest: classification
##                        Number of trees: 500
## No. of variables tried at each split: 3
##
##          OOB estimate of  error rate: 9.45%
## Confusion matrix:
##         no yes class.error
## no    2061  34  0.01622912
## yes    191  94  0.67017544
```

In the output we see that 500 trees were fitted and 3 variables were tried at each split. One feature of random forests is that there is an inbuilt kind of validation of the predictions made. Since bootstrap samples are used to grow trees, some samples are out of bag and are not used. In the output we see that the out of bag error rate is 9.45%. It is far easier to fit a model to in sample data; the more difficult thing is for a model to fit data out of sample.

It is the emphasis on predictive accuracy that is the hallmark of random forests and related methods. Breiman writes:

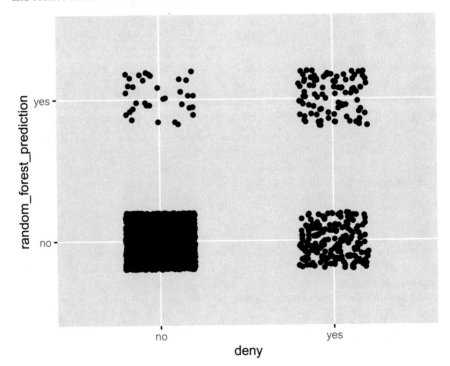

Fig. 16.8 Scatter plot of predictions from random forest versus deny. Note that data have been jittered

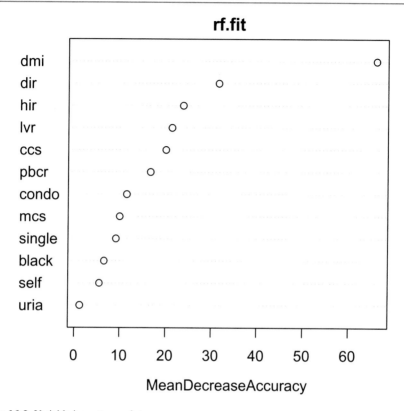

Fig. 16.9 Variable importance plot

The most obvious way to see how well the model box emulates nature's box is this: put a
case x down nature's box getting an output y. Similarly, put the same case x down the model
box getting an output y'. The closeness of y and y' is a measure of how good the emulation
is. For a data model, this translates as: fit the parameters in your model by using the data,
then, using the model, predict the data and see how good the prediction is.

We can get predictions from the random forest with:

```
hdma$random_forest_prediction <- predict(rf.fit)
```

The data itself contains the variable deny, i.e. this is output from nature's box (in
Breiman's words in the quote above). We make a scatter plot of the predictions from
the random forest versus the output from nature's box (Fig. 16.8). Figure 16.8 is the
graphical counterpart of the Confusion Matrix reported in the random forest output
above.

```
ggplot(hdma, aes(y = random_forest_prediction, x = deny)) +
  geom_jitter(height = 0.2, width = 0.2)
```

Varian (2014, p. 15) writes, 'One defect of random forests is that they are a bit of a black box–they don't offer simple summaries of relationships in the data. …However, random forests can determine which variables are "important" in predictions in the sense of contributing the biggest improvements in prediction accuracy'.

```
varImpPlot(rf.fit, type = 1)
```

We get measures of variable importance in Fig. 16.9 based on the mean decrease of accuracy in predictions on the out of bag samples when a given variable is excluded from the model. The three most important variables are dmi, dir and hir.

16.5   Resources

Varian (2014) is a nice paper, introducing machine learning and covering trees and random forests, and motivating the economist to learn the tricks of machine learning. James et al. (2013) explain trees and random forests very clearly. The pdf of their book is downloadable from the book website.

References

Berk, R.A. 2008. *Statistical learning from a regression perspective*. New York: Springer.

Biau, G., and E. Scornet. 2016. A random forest guided tour. *TEST* 25 (2): 197–227. https://doi.org/10.1007/s11749-016-0481-7.

Breheny, P., and W. Burchett. 2017. Visualization of regression models using visreg. *The R Journal* 9: 56–71.

Breiman, L. 2001. Statistical modeling: The two cultures. *Statistical Science* 16 (3): 199–231.

Gelman A., and J. Hill. 2006. *Data Analysis Using Regression and Multilevel/Hierarchical Models*. New York: Cambridge University Press.

James, W.D., T. Hastie, and R. Tibshirani. 2013. *An introduction to statistical learning with applications in R*. New York: Springer.

Milborrow, S. 2019. rpart.plot: Plot 'rpart' models: An enhanced version of 'plot.rpart'. *R package version 3.0.8*. https://CRAN.R-project.org/package=rpart.plot.

Therneau T, and B. Atkinson. (2019) rpart: Recursive partitioning and regression trees. *R package version 4.1-15*. https://CRAN.R-project.org/package=rpart.

Varian, H.R. 2014. Big data: New tricks for econometrics. *Journal of Economic Perspectives* 28 (2): 3–28.